Studies in Logic
Volume 101

The Logic of Partitions
With Two Major Applications

Studies in Logic Series Editor
Dov Gabbay   dov.gabbay@kcl.ac.uk

The Logic of Partitions
With Two Major Applications

David Ellerman

ISBN 978-1-84890-440-8

College Publications
Scientific Director: Dov Gabbay
Managing Director: Jane Spurr

http://www.collegepublications.co.uk

To the memory of Larry Harper,
combinatorial mathematician,
close associate of Gian-Carlo Rota,
and long-time friend.

Preface

My PhD. work in the late 1960's was in mathematical logic and category theory (later absorbed into topos theory). I was the first doctoral student of Rohit Parikh who taught me mathematical logic and to think rigorously. But my real formation as a mathematician was in collaboration in the early 1980's with Gian-Carlo Rota. We wrote a joint paper that took my Erdös number from infinity down to 3. But most of my employment in later years was in Economics, not mathematics or logic. A big shock came in 1999 just before my retirement when Rota died relatively young at age 66.

By 2005, I became so sick of the ideological prejudices in Economics, that I decided at age 62 to go back into mathematical research. It was an interesting transformation. After not doing any serious mathematics for decades, it took about a year of "rewiring synapses" to begin to think again like a mathematician. But it was a different kind of mathematical thinking. Young mathematicians think that serious work requires going out to the frontiers and then doing some notable work. I had a different mentality. I felt that there might be new theories, bypassed by those rushing to the frontiers of existing math, that were like fruit low-hanging enough that could be developed by a retooled mathematician in his sixties.

The other important influence was Rota's tragic death a few years earlier and the huge body of his work that was only partially developed. Some of us who had worked with him wanted to pick up some of his unfinished strands to further develop. One of those strands was a 1996 paper by Rota and some colleagues entitled: "Logic of Commuting Equivalence Relations." Rota was well aware of the category-theoretic duality between subsets and partitions (or equivalence relations). Since ordinary logic starts with the Boolean logic of subsets (usually presented in the special case of "propositional logic"), Rota had the idea of developing a logic of that dual concept, a logic of partitions or equivalence relations. But the use of the word "logic" in that 1996 paper was somewhat ambitious since there was no known implication operation on equivalence relations or partitions, only the lattice operations of join and meet that had been known in the nineteenth century (e.g., Dedekind and Schröder). Furthermore, it was then known that partitions were so versatile that any formula in the language of lattice theory that held for all lattices of partitions would be a lattice theory identity. Hence Rota considered only a special type of equivalence relation called "commuting" that was previously developed by Marie-Louise Dubreil-Jacotin and her husband, Paul Dubreil, in a 1939 paper "Théorie algébrique des relations d'équivalence." Rota was fascinated by commuting equivalence relations which played a significant role in his MIT lectures on combinatorial theory. But the idea of a logic of general equivalence relations or partitions went no further.

In fact, no new operations of partitions, such as the implication operation, were developed throughout the twentieth century. There were a number of reasons why the general logic of partitions did not develop in the twentieth century. Even though the duality of subsets and partitions was known at least since the development of category theory in the mid-twentieth century, the exposition of mathematical logic almost always began with the special case of the logic of propositions rather than the logic of subsets–as one can verify by consulting yesterday's and today's logic textbooks. Since propositions do not have a category theoretic dual, the whole idea of a dual logic of partitions was not "in the air."

The second hindrance was the "lattice of partitions" that was presented in the literature (e.g., Birkhoff's book *Lattice Theory*) was typically the lattice of equivalence relations with the universal equivalence relation $U \times U$ as the top and using the partial order of inclusion–which was also the lattice structure used in Rota's paper. Yet, it turns out that the implication operation $\sigma \Rightarrow \pi$ on partitions and the modus ponens validity, $(\sigma \wedge (\sigma \Rightarrow \pi)) \Rightarrow \pi$, only have that form if the lattice is turned upside down to form the genuine lattice of partitions with the partial order of refinement. In terms of equivalence relations, the modus ponens validity has the form $\pi - (\sigma \vee (\pi - \sigma))$, which would not strike logicians as the modus ponens formula (disguised as a relationship between equivalence relations that would always evaluate to the top $U \times U$ regardless of what equivalence relations on U were substituted for σ and π). In my work on partition logic at the University of California Riverside, I was encouraged by combinatorial mathematics professor Larry Harper who had an earlier and more in-depth collaboration with Gian-Carlo Rota. In particular, Larry agreed that the lattice of equivalence relations should be turned over to form the lattice of partitions and that got me on the right track.

Eventually, perhaps with a little luck, I was able to define the implication operation on partitions. Soon it became clear that there were, in fact, two algorithms that could be used to define all the Boolean operations on partitions. That was the beginnings of the logic of partitions *per se* that is developed in this book. Much of the development was just picking low-hanging fruit using analogies with the logic of subsets–as well as helpful analogies with the logic of open subsets of a topological space (Heyting algebras). But partitions are much more complicated than the dual subsets, so partition logic is corresponding considerably more complex than the logic of subsets. Hence this book is not a monograph giving a complete development of partition logic–as one might have for subset logic. It is "early days" for partition logic so this book should be seen only as an early accumulation of some results that do not include, for instance, a Hilbert-style axiom system for the logic.

Nevertheless, it is already clear that partition logic has some major applications. The first major application is again foreshadowed by Rota who emphasized the analogy: $\frac{\text{Probability}}{\text{Subsets}} \approx \frac{\text{Information}}{\text{Partitions}}$. Boole logically developed finite probability theory starting as the quantitative notion of subsets, e.g., the probability of getting a subset S of a finite equiprobable sample space U is just the normalized number of elements: $\Pr(S) = \frac{|S|}{|U|}$. Hence using Rota's analogy, the logical notion of information should start with the quantitative notion of the "size" of a partition. But one needs to turn the lattice of equivalence relations upside down to see the proper notion of size of a partition. An ordered pair (u, u') in an equivalence relation $E \subseteq U \times U$ is a pair of elements of $U = \{u_1, ..., u_n\}$ in the same block or equivalence class of the equivalence relation E. Those ordered pairs are called the *indistinctions* or *indits* of the corresponding partition. The complement $U \times U - E$ (sometimes called an "apartness relation") consists of ordered pairs $(u, u') \in U \times U$ where u and u' are in different equivalence classes, i.e., different blocks of the corresponding partition. Those ordered pairs are called the *distinctions* or *dits* of the partition so the complement $U \times U - E$ is just the set of dits, the ditset, of the corresponding partition.

When the lattice is turned over, then the partial order is turned from the inclusion relation between equivalence relations, i.e., inditsets, to the inclusion relation between ditsets. The partial order of refinement, written $\sigma \precsim \pi$, is usually defined as holding when for each block B of π, there is a block C of σ such that $B \subseteq C$. But that is equivalent to the inclusion partial order on ditsets, i.e., $\sigma \precsim \pi$ if and only if $\text{dit}(\sigma) \subseteq \text{dit}(\pi)$. Since the partial order in the Boolean lattice of subsets is the inclusion relation between subsets, it is clear that the analogy is between *elements* of a subset and *distinctions* of a partition. Hence the subset-partition duality can be refined into the duality of the elements of a subset and the distinctions of a partition. Then we can finally answer the question raised by Rota's analogy; the quantitative notion of a partition is the size of its ditset. The basic logical notion of information in a partition is just the normalized size of its ditset, so the initial definition of *logical entropy* for $\pi = \{B_1, ..., B_m\}$ is:

$$h(\pi) = \frac{|\text{dit}(\pi)|}{|U \times U|} = \frac{|U \times U - \cup_j (B_j \times B_j)|}{|U \times U|} = 1 - \sum_j \left(\frac{|B_j|}{|U|}\right)^2 = 1 - \sum_j \Pr(B_j)^2 = \sum_{j \neq k} \Pr(B_j) \Pr(B_k)$$

with equiprobable points in U. If there is a general probability distribution $p = (p_1, ..., p_n)$ on the points of U, then the logical entropy of π is just the value of product probability measure $p \times p$ on the ditset $\text{dit}(\pi) \subseteq U \times U$. The interpretations of $\Pr(S)$ and $h(\pi)$ are thus also analogous. One random draw from U gets an element of S with the probability $\Pr(S)$, and two random draws from U gets a distinction of π with the probability $h(\pi)$. This also (finally) gives a logical definition of information, i.e., information as distinctions.

The Venn-diagram compound notions of joint, conditional, and mutual logical entropy are just the values of that probability measure on the corresponding sets, $\text{dit}(\pi) \cup \text{dit}(\sigma)$, $\text{dit}(\pi) - \text{dit}(\sigma)$, and $\text{dit}(\pi) \cap \text{dit}(\sigma)$. The usual notion of entropy, Shannon entropy, is defined so that the usual Venn diagrams relationships hold even though Shannon entropy is not a measure in the sense of measure theory. But there is a monotonic dit-bit transform that converts all the compound notions of logical entropy into the corresponding notions of Shannon entropy and that transform preserves Venn diagram relationships.

Hence the first major application of partition logic is simply its quantitative version as a logical definition of information analogous to the way Boole started probability theory as a quantitative version of subset logic. That supplies a much-needed *logical* foundation for information theory developed in my 2021 book *New Foundations for Information Theory: Logical Entropy and Shannon Entropy*.

The second major application was to the century-old problem of understanding the reality that quantum mechanics (QM) describes so well. QM was consolidated in the mid-1920's and over the last century, there has been no agreement on the nature of reality at the quantum level. New so-called "interpretations" are continually being created without any noticeable convergence. Otherwise sane physicists are driven to rather crazy ideas, e.g., the many-worlds interpretation, when confronted with the "paradoxes" of quantum theory. It is in this intellectual "demolition derby" of quantum interpretations where partition logic and logical entropy offer a new approach to corroborate an interpretation promoted by Werner Heisenberg and Abner Shimony among others. This new approach "cuts at the joint" between the mathematics and the physics of quantum mechanics. The mathematics is quite distinctive and different from the mathematical framework of classical physics. The new approach asks: "Where does the distinctive mathematics of QM come from?". The answer is that the math of QM is the vector-space or, particularly, Hilbert-space version of the mathematics of partitions. The argument is based, in part, on using a semi-algorithmic procedure to build a translation-dictionary between set-level partition math and Hilbert-space QM math. The next step in the argument is to ask: "What basic concepts are represented at the logical level by partitions?". The answer is the concepts variously described as distinctions versus indistinctions, definiteness versus indefiniteness, distinguishability versus indistinguishability, or difference versus identity (i.e., inequivalence versus equivalence in terms of equivalence relations).

Then we look at quantum theory and ask: "What is the essential non-classical concept in QM?". The answer is the notion of superposition. But that non-classical notion has been known from the beginning so, "Why has there been so little progress in understanding the reality behind the notion of superposition?". One answer might be the lack of development in partition math in the twentieth century. But a better answer lies in the mathematics of QM itself. A Hilbert space is a vector space over the complex numbers $\mathbb{C}$, and the complex numbers are the natural mathematics to describe waves (the polar representation of a complex number is an amplitude and phase of a wave). Indeed, QM is often called "wave mechanics" and the "wave function" is a commonly used mathematical tool to represent the quantum state. Hence superposition has usually been interpreted as simply like the addition of waves–just as water waves might add and interfere with each other. But after a century of looking, there are no physical waves found at the quantum level (much to the dismay of Erwin Schrödinger who invented the eponymous wave equation)–only what might be called "probability waves" which are not physical entities. The quantum world is not wave-world. Indeed, the math of QM is formulated using the complex numbers for reasons that have nothing to do with waves, namely that the complex numbers are the algebraically complete extension of the real numbers so that the real-valued quantum observables will then have a complete set of eigenvectors. In other words, the whole wave interpretation of QM math was mistakenly giving an ontological importance to the wave-like computational artifacts present in any vector space over the complex numbers.

How to escape this conundrum that the wave-like math is not reflected ontologically in a quantum level wave-world? What is needed is a totally different interpretation of superposition. And *that* interpretation is supplied by the mathematics of partitions. At the simple logical level, a partition is made up of blocks of elements of the underlying set. Each block is an equivalence class that says according to this partition, these elements in the block are equated or blobbed together with no distinctions between them–since the distinctions are between different blocks. Thus the blocks with two or more elements are the logical version of a superposition of eigenstates in QM math. The block is indefinite or indistinct on the differences between elements (or eigenstates). The elements in the set (or eigenstates in the QM version) represent (not different particles but) different states of a particle that are equated, blobbed together, or cohered together in the superposition. That is the non-wave reinterpretation of superposition that corroborates an interpretation of QM proposed by Heisenberg, Shimony, and others. The key idea behind this version of superposition is that a particle can be in an objectively indefinite state like a particle in a superposition state of "here" and "there", i.e., it is "not definitely here and not definitely there, but definitely not anywhere else." Partitions have nothing to do with waves so this partition approach to the math of QM is key to making the transition from wave-world to indefinite-world.

While constantly using the wave-math without finding any physical waves, most quantum theorists do recognize the reality of indefinite states. That is, when the quantum state is a superposition in the basis of the observable being measured, then it is widely recognized that the real quantum state does not have a definite value before the measurement which causes the quantum jump into a definite state. And the set-level version of (projective) measurement is just the partition join operation from partition logic. Heisenberg, Shimony, and others then extrapolate that notion of an indefinite state to the whole of quantum-level reality. The quantum world is indefinite-world, not wave-world. And the set-level mathematics to represent definiteness versus indefiniteness is the math of partitions with the math of QM being the Hilbert space version of that partition math. That objective indefiniteness interpretation of QM is the second major application of partition logic. It is further developed in my 2023 book *Partitions, Indefiniteness, and Quantum Reality: The Objective Indefiniteness Interpretation of Quantum Mechanics.*

Contents

Chapter 1

The logical operations on set partitions

1.1 Introduction to partitions

1.1.1 The two mathematical logics of subsets and partitions

There are fundamentally two mathematical logics. One the Boolean logic of subsets [13], usually presented today in the special case of propositional logic, which has many sublogics and extensions, the most important being the intuitionistic logic usually modeled by the open subsets of a topological space. The other co-fundamental mathematical logic is the topic of this book, the logic of partitions. We are using "logic" in a mathematical sense as being *about* basic mathematical objects, subsets of a universe set or partitions on a universe set.[1] Logic, in this mathematical sense, is not *about* propositions, although, as with any mathematical theory, it involves propositions about the basic objects, e.g., that an element is in a subset or that a distinction is made by a partition. Moreover, by taking the universe set to be the one element set 1, with two subsets $\emptyset$ and 1, there is a special case of subset logic, namely propositional logic, that can be interpreted as being about propositions with $\emptyset$ representing falsehood and 1 representing truth.

In the nineteenth century, what is now called "propositional logic," was developed as the logic of subsets.

The algebra of logic has its beginning in 1847, in the publications

---

[1] We are not using the word "logic" for any syntactic axiom system using logical connectives, but as a theory about certain fundamental mathematical notions. Indeed, today a Hilbert-style axiom system for partition logic has yet to be developed.

of Boole and De Morgan. This concerned itself at first with an algebra or calculus of classes, to which a similar algebra of relations was later added. Though it was foreshadowed in Boole's treatment of "Secondary Propositions," a true propositional calculus perhaps first appeared from this point of view in the work of Hugh MacColl, beginning in 1877. [17, pp. 155-56]

Today the original subset version of propositional logic seems to be most often noted in the context of the category-theoretic treatment.

The propositional calculus considers "Propositions" p, q, r,... combined under the operations "and", "or", "implies", and "not", often written as $p \wedge q$, $p \vee q$, $p \Rightarrow q$, and $\neg p$. Alternatively, if P, Q, R,... are subsets of some fixed set U with elements u, each proposition p may be replaced by the proposition $u \in P$ for some subset $P \subset U$; the propositional connectives above then become operations on subsets; intersection $\wedge$, union $\vee$, implication ($P \Rightarrow Q$ is $\neg P \vee Q$), and complement of subsets.[74, p. 48]

Why should partition logic be paired with subset logic as the two fundamental mathematical logics? The answer to that question awaited the development of category theory and the associated notion of duality developed in the 1940's ([23]; [72])–although it was foreshadowed in the parallelism or 'duality' between the subobjects (subsets, subgroups,...) and quotient objects (quotient sets, quotient groups,...) in pre-category-theoretic abstract algebra. The category-theoretic dual to a subset or more generally a subobject or 'part' is a quotient set (or equivalently, an equivalence relation or partition) or more generally a quotient object. "The dual notion (obtained by reversing the arrows) of 'part' is the notion of partition." [69, p. 85] F. William Lawvere and Robert Rosebrugh go on to treat "Logic as the Algebra of Parts" [69, p. 193] but do not suggest a dual logic as the algebra of partitions.

Partition logic is at the same mathematical level as subset logic since the semantic models for both are constructed from (partitions on or subsets of) arbitrary unstructured universe sets with no topologies, no ordering relations, and no other structure on the sets.

Since the Boolean logic of subsets was known in the nineteenth century and category-theoretic duality was known since the middle of the twentieth century, why did it take so long for the logic of partitions to be developed? ([24]; [25]) One partial answer is simply the almost exclusive presentation of Boolean logic as propositional logic, and "propositions" do not have a mathematical dual (only a complement). The basic object of Boolean logic, i.e., subsets, needed to presented at the right level of generality in order for the dual notion of partitions to be clear. Also it should be noted that although subsets and partitions are at the same level of mathematical fundamentality, partitions are more complicated than subsets and that is reflected is the considerably higher level of complication in partition logic.

Another partial answer was the lack to the logical operations on partitions beyond the join and meet. Those lattice-theoretic operations on partitions were known in the nineteenth century (e.g., Richard Dedekind and Ernest Schröder), but a logic to compare to subset logic awaited at least the operation of implication on partitions. Yet, throughout the twentieth century, the only operations on partitions that were studied were the lattice operations of join and meet. In a 2001 paper commemorating Gian-Carlo Rota, the three authors first note the fundamentality of partitions and then acknowledge the sole operations of join and meet.

> Equivalence relations are so ubiquitous in everyday life that we often forget about their proactive existence. Much is still unknown about equivalence relations. Were this situation remedied, the theory of equivalence relations could initiate a chain reaction generating new insights and discoveries in many fields dependent upon it.
>
> This paper springs from a simple acknowledgement: the only operations on the family of equivalence relations fully studied, understood and deployed are the binary join ∨ and meet ∧ operations.[15, p. 445]

Gian-Carlo Rota indeed had the idea of developing a logic of equivalence relations or partitions, but without the implication operation, the only identities would be those formulas that hold on all lattices of partitions. Moreover, it was known that partitions are so versatile that the only identities that hold on all lattices of partitions or equivalence relations are the general lattice-theoretic identities [112]. Hence without the implication or other logical operations on partitions, the only way to develop a specific logic of equivalence relations was to focus on a specific type such as commuting equivalence relations [22], and that is what Rota and colleagues did [41].

1.1.2 The duality of elements and distinctions

The duality between subsets and partitions can be traced back to a more basic duality between the elements of a subset and the distinctions of a partition ('its' and 'dits'). In the category *Set* of sets and functions, the objects are sets and the morphisms are functions between sets. The notion of a function is naturally defined using the dual notions of elements and distinctions.

Given two sets X and Y, consider a binary relation $R \subseteq X \times Y$.

The relation R is said to *transmit elements* if for all $x \in X$, there is an ordered pair $(x, y) \in R$ for some $y \in Y$.

The relation R is said to *reflect elements* if for all $y \in Y$, there is an ordered pair $(x, y) \in R$ for some $x \in X$.

The relation R is said to *transmit distinctions* if for any $(x, y) \in R$ and $(x', y') \in R$, if $x \neq x'$, then $y \neq y'$.

The relation R is said to *reflect distinctions* if for any $(x, y) \in R$ and $(x', y') \in R$, if $y \neq y'$, then $x \neq x'$.

It might be noted that the definitions of "reflect" and "transmit" just interchange the roles of X and Y. Then a binary relation R is said to be *functional* or to define a *function* if it is defined everywhere on X, i.e., transmits elements, and if it is single-valued, i.e., reflects distinctions. The dual "turn-arround-the-arrows" notion of a morphism in the opposite category Set^{op} is obtained by interchanging elements and distinctions, i.e., a *cofunction* is a binary relation R that transmits distinctions and reflects elements. In this manner, the notion of duality in *Set*, that provides the duality between subsets and partitions, can be traced back to elements and distinctions [31]. It might also be noted that when R transmits elements and reflects distinctions so that it is a function $f : X \to Y$, then the two special types of functions, injective (one-to-one) and surjective (onto) are defined respectively as transmitting distinctions and reflecting elements. Each function $f : X \to Y$ has an associated subset, namely the image $f(X) \subseteq Y$, and an associated partition, namely the inverse-image or coimage (or kernel) $\{f^{-1}(y)\}_{y \in f(X)}$ as shown in Figure 1.1.

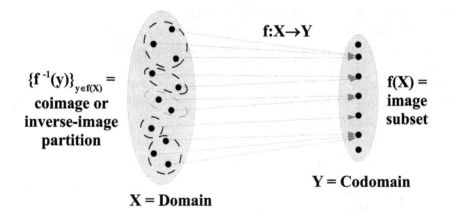

$$\{f^{-1}(y)\}_{y \in f(X)} =$$
coimage or
inverse-image
partition

f:X→Y

$$f(X) =$$
image
subset

Y = Codomain

X = Domain

Figure 1.1: Image of f is a subset of the codomain of f; inverse-image of f is a partition on the domain of f

1.1.3 Partitions and equivalence relations

A *partition* $\pi = \{..., B, ..., B', ...\}$ on a universe set U (arbitrary cardinality unless otherwise specified) is a set of non-empty subsets, called *blocks* (or sometimes *cells*), $B, B', ...$ of U that are pairwise disjoint and whose union is U. An alternative definition is that a partition $\pi = \{..., B, ..., B', ...\}$ is a set of subsets of U such that every subset $S \subseteq U$ can be uniquely represented as a union of subsets of the blocks. If the union of the blocks did not exhaust U, then the difference $U - \cup_{B \in \pi} B \neq \emptyset$ would not be represented by a union of subsets of the blocks. And if any $B \cap B' \neq \emptyset$, then that non-empty subset could be represented in two ways by subsets of the blocks.[2]

A *distinction* or *dit* of a partition π is an *ordered* pair of elements (u, u') in different blocks of the partition, while an *indistinction* or *indit* of π is an ordered pair of elements (u, u') in the same block. The *ditset* $\mathrm{dit}\,(\pi) \subseteq U \times U$ of π is the sets of distinctions, and the inditset $\mathrm{indit}\,(\pi) = U \times U - \mathrm{dit}\,(\pi)$ is the set of indistinctions of π as illustrated in Figure 1.2:

$$\mathrm{indit}(\pi) = \cup_{B \in \pi} B \times B \text{ and } \mathrm{dit}\,(\pi) = \cup_{B, B' \in \pi, B \neq B'} B \times B' = U \times U - \mathrm{indit}\,(\pi).$$

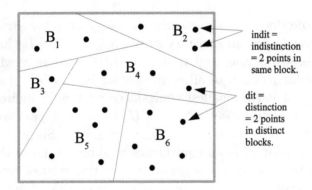

Figure 1.2: Distinctions and indistinctions of a partition

---

[2]The major application of the mathematics of partitions is to supply the interpretation of the math of quantum mechanics in terms of objective indefiniteness (see below). Part of that demonstration is a linearization method that linearizes a set partition to a direct-sum decomposition (DSD) of a vector space–which is defined as a set of subspaces such that any vector in the space can be uniquely represented as a sum of vectors from the subspaces in the DSD. That is the source of the alternative definition of a set partition.

An *equivalence relation* $E \subseteq U \times U$ on U is a binary relation that is reflexive (for all $u \in U$, $(u,u) \in E$), symmetric (for any $u, u' \in U$, if $(u, u') \in E$, then $(u', u) \in E$), and transitive (for any $u, u', u'' \in U$, if $(u, u') \in E$ and $(u', u'') \in E$, then $(u, u'') \in E$). As a binary relation indit $(\pi) \subseteq U \times U$ on U, the inditset is the equivalence relation associated with π, and the ditset dit $(\pi) = U \times U -$ indit (π) is its complement. A *partition relation* R on U (also called an "apartness relation") is the complement of an equivalence relation in $U \times U$ so it is irreflexive (i.e., $(u, u) \notin R$ for all $u \in U$), symmetric, and intransitive (i.e., for any $(u, u'') \in R$ and any $u' \in U$ either $(u, u') \in R$ or $(u', u'') \in R$, or, in other words, $U \times U - R$ is transitive). Every equivalence relation on U is the inditset of a partition on U (take the blocks as the equivalence classes) and vice-versa. Every partition relation on U is the ditset of a partition on U and vice-versa.

There is a suggestive analogy between equivalence and partition relations and the closed and open subsets of a topological space T. The complement of a closed set of T is an open set, and the intersection of an arbitrary number of closed subsets is closed. Given an arbitrary subset $S \subseteq T$, the topology on T defines the closure $\overline{S}$ (i.e., the intersection of all closed subsets containing S) which is the smallest closed set containing S. And given S, there is the largest open subset int $(S) \subseteq S$ contained in S which is obtained by the complement of the closure of the complement, i.e., int $(S) = \left(\overline{S^c}\right)^c$ (where T^c is the complement of a subset T).

Partition logic does not assume any topology or other structure on the universe U. Nevertheless, an arbitrary subset $S \subseteq U \times U = U^2$ has a naturally defined reflexive, symmetric, and transitive (RST) closure $\overline{S}$ which could be obtained as the intersection of all equivalence relations ($\approx$ "closed subsets") containing S. That is the smallest equivalence relation or inditset containing S. Similarly, given any subset $S \subseteq U \times U$, the largest partition relation or ditset contained in S is obtained as int $(S) = \left(\overline{S^c}\right)^c$. Since the intersection of an equivalence relation is an equivalence relation, we know that the union of partition relations or ditsets is always a partition relation or ditset. Thus int (S) can also be obtained as the union of all the partition relations contained in S. Since int $(S) = \left(\overline{S^c}\right)^c = \text{dit}(\pi)$ for some partition π on U, we also have: indit$(\pi) = \overline{S^c}$.

But the analogy is not complete; otherwise partition logic would just be a special case of the intuitionistic logic of open subsets. In particular, the RST closure operation is not topological in the sense that the union of RST-closed subsets of $U \times U$, i.e., the union of equivalence relations, is not necessarily a RST-closed subset (an equivalence relation). As will be later seen, there are valid formulas of partition logic that are not valid in intuitionist logic and there are intuitionistic validities that are not valid in partition logic–although the valid formulas of both these logics are properly contained in the classical tautologies of subset logic.

1.2 The join and meet operations on partitions

1.2.1 The set-of-blocks definitions of join and meet

The lattice operations of join and meet of partitions will first be defined in the traditional (i.e., nineteenth century) way in terms of their blocks, and then in some equivalent new ways that generalize easily to the other logical operations. Let $\Pi(U)$ be the set of all partitions on U.[3]

Given two partitions $\pi = \{..., B, ...\}$ and $\sigma = \{..., C, ...\}$ in $\Pi(U)$, the *join* $\pi \vee \sigma$ is the partition whose blocks are the non-empty intersections $B \cap C$ for $B \in \pi$ and $C \in \sigma$. The union of two ditsets is a ditset, and indeed:

$$\text{dit}(\pi) \cup \text{dit}(\sigma) = \text{dit}(\pi \vee \sigma).$$

The *meet* of π and σ, written $\pi \wedge \sigma$, may be defined using the notion of two blocks $B \in \pi$ and $C \in \sigma$ having a non-empty intersection, written $B \between C$ (following Ore [79]). If two blocks intersect, think of them as blobbing together like two touching drops of a liquid. Then we could have a sequence $B \between C \between B' \between C' \between ...$ and the blocks of the meet are the minimal unions of such sets of overlapping blocks that have no overlaps with blocks outside the union. Hence the blocks of the meet are the smallest subsets of U that are a union of some blocks of π and also a union of some blocks of σ. It was previously noted that the RST-closure operation was not topological in the sense that the union of two inditsets is not necessarily an inditset. Hence the intersection of two ditsets is not necessarily a ditset, but we can take the interior $\text{int}[\text{dit}(\pi) \cap \text{dit}(\sigma)]$. To prove that is the ditset of the meet, consider two elements u and u' as being directly equated, $u \sim u'$ if u and u' are in the same block of π or σ so the set of directly equated pairs is: $\text{indit}(\sigma) \cup \text{indit}(\pi)$. Then u and u^* are in the same block of the meet in $\Pi(U)$ if there is a finite

---

[3]The lattice of subsets is degenerate if the top equals the bottom as in the case of $\wp(\emptyset)$ so it is commonly assumed that $|U| \geq 1$ when working with the lattice or Boolean algebra of subsets of U. In the partition case, there two degenerate cases where the top equals the bottom, namely for $U = \emptyset$ or $U = 1$ (the one element set). The partition $\{1\}$ is the inverse-image of any function $1 \to Y$ and the empty partition $\emptyset$ is the inverse-image of the unique empty function $\emptyset \to Y$. Hence, unless otherwise stated, we will assume $|U| \geq 2$.

sequence $u = u_1 \sim u_2 \sim \ldots \sim u_n = u^*$ that indirectly equates u and u^* by a sequence $B \between C \between B' \between C' \between \ldots$. The operation of indirectly equating two elements is just the closure operation (transitivity) in the closure space so the set of pairs indirectly equated, i.e., equated in the meet $\sigma \wedge \pi$ in $\Pi(U)$, is:

$$\text{indit}(\sigma \wedge \pi) = \overline{(\text{indit}(\sigma) \cup \text{indit}(\pi))}.$$

The complementary subset of $U \times U$ is the dit set of the meet of the partitions:

$$\text{dit}(\sigma \wedge \pi) = \text{indit}(\sigma \wedge \pi)^c = \overline{(\text{indit}(\sigma) \cup \text{indit}(\pi))}^c = \text{int}(\text{dit}(\sigma) \cap \text{dit}(\pi)).$$

The partial order on the set of partitions $\Pi(U)$ on U that makes these operations in the least upper bound and greatest lower bound respectively is the refinement partial ordering: σ is *refined by* π or π *refines* σ, written $\sigma \precsim \pi$, if for every $B \in \pi$, there is a block $C \in \sigma$ such that $B \subseteq C$. In terms of functions, π and σ define the canonical surjections $U \to \pi$ and $U \to \sigma$ (take each element of U to its block in π or σ), $\sigma \precsim \pi$ if and only if (iff) there exists a function $\pi \to \sigma$ to make the following triangle commute, i.e., $U \to \pi \to \sigma = U \to \sigma$.

$$
\begin{array}{ccc}
U & \longrightarrow & \pi \\
\downarrow & {\scriptstyle \exists} \swarrow & \\
\sigma & &
\end{array}
$$

Refinement is a partial ordering in the sense that it is reflexive, transitive, and anti-symmetric (i.e., if $\sigma \precsim \pi$ and $\pi \precsim \sigma$, then $\sigma = \pi$). Moreover, the refinement relation on partitions is just the inclusion partial order on their ditsets (or partition relations):

$$\sigma \precsim \pi \text{ iff } \text{dit}(\sigma) \subseteq \text{dit}(\pi).$$

Then it follows from $\text{dit}(\sigma) \cup \text{dit}(\pi) = \text{dit}(\pi \vee \sigma)$ that the join $\pi \vee \sigma$ is the least upper bound on π and σ. Similarly, it follows from $\text{int}[\text{dit}(\pi) \cap \text{dit}(\sigma)] = \text{dit}(\pi \wedge \sigma)$ that the meet is the greatest lower bound on π and σ. It should be noted that many of older texts ([10]; [49]) actually deal with the lattice of equivalence relations with the partial order of inclusion even thought it may be called the "lattice of partitions." Since that is the opposite partial order, the join and meet operations are interchanged.

The join and meet operations on partitions make $\Pi(U)$ into a lattice. Moreover, the lattice $\Pi(U)$ always has a maximum element or top $\mathbf{1}_U = \{\{u\}\}_{u \in U}$, called the *discrete partition*, whose blocks are the singletons of the elements of U so that $\text{dit}(\mathbf{1}_U) = U \times U - \Delta$ where $\Delta = \text{indit}(\mathbf{1}_U)$ is the diagonal set of self-pairs (u, u) for $u \in U$. And the lattice also has a minimum partition or bottom $\mathbf{0}_U = \{U\}$, called the *indiscrete partition*,[4] with only one block U so that dit $(\mathbf{0}_U) = \emptyset$ and indit $(\mathbf{0}_U) = U \times U$.

Let $\mathcal{O}(U \times U)$ be the set of ditsets (partition relations) as the 'open' subsets of $U \times U$ ordered by inclusion and with the join being union and the meet being the interior of the intersection. The result is a lattice isomorphic to the lattice of partitions $\Pi(U)$ under the mapping $\pi \longmapsto$ dit (π) so the lattice of partitions is represented by the 'open' subsets of $U \times U$:

$$\Pi(U) \cong \mathcal{O}(U \times U),$$

The dual counterpart to the lattice of partitions $\Pi(U)$ on U is the lattice of subsets $\wp(U)$ of U. The duality carries through to the elements and distinctions on each side as indicated in Table 1.1.

Its & Dits	Lattice $\wp(U)$ of subsets of U	Lattice of partitions $\Pi(U)$ on U
Its or Dits	Elements of subsets	Distinctions of partitions
Partial order	Inclusion of subsets	Inclusion of ditsets
Join	Union of subsets	Union of ditsets
Meet	Subset of common elements	Ditset of common dits
Top	Subset U with all elements	Partition $\mathbf{1}_U$ with all distinctions
Bottom	Subset $\emptyset$ with no elements	Partition $\mathbf{0}_U$ with no distinctions

Table 1.1: Elements and Distinctions (Its & Dits) duality between two lattices

1.2.2 The ditset definitions of join and meet

The previous definitions of the join and meet were given in terms of the blocks of the partitions so they might be called the "set-of-blocks" definitions. But then it was noted that dit $(\pi \vee \sigma) =$ dit $(\pi) \cup$ dit (σ) and dit $(\pi \wedge \sigma) =$ int $[$dit $(\pi) \cap$ dit $(\sigma)]$. That means that we can define the join and meet in terms of the ditsets and the interior operation. That gives us the ditset definitions of the lattice operations.

To see the equivalence in the join definitions, if (u, u') is in one of the ditsets, say dit (π), then $u \in B$ and $u' \in B'$ for some $B \neq B'$, so $u \in B \cap C$ for some $C \in \sigma$ and $u' \in B' \cap C'$ for some $C' \in \sigma$ (even when $C = C'$) so (u, u') is a dit of the set-of-blocks defined $\pi \vee \sigma$. Conversely if (u, u') is a dit of the set-of-blocks definition, i.e., $u \in B \cap C$ and $u' \in B' \cap C'$, and even if $B = B'$ or $C = C'$ but not both, then (u, u') is in one of the ditsets.

---

[4]It is nicknamed "the Blob" since as in the eponymous Hollywood movie, it absorbs everything it meets: $\mathbf{0}_U \wedge \pi = \mathbf{0}_U$.

The ditset definition of the meet $\pi \wedge \sigma$ immediately establishes it as the greatest lower bound on π and σ so to prove the equivalence with the set-of-blocks definition, it suffices to show that the set-of-blocks approach also gives the greatest lower bound. Consider any lower bound $\tau = \{..., D, ..., D', ...\}$ such that $\tau \precsim \pi, \sigma$. Then for any $B \in \pi$, there is a $D \in \tau$ with $B \subseteq D$ and for any $C \in \sigma$, there is a $D' \in \tau$ with $C \subseteq D'$. But consider any block in $\pi \wedge \sigma$ which is an exact union of blocks of π and at the same time an exact union of blocks of σ. Consider any B contained in that meet-block. It is contained in a block $D \in \tau$, but B intersects some $C \in \sigma$, i.e., $B \between C$, and thus $B \cap C \subseteq D$. But that means that the block of τ that C is contained in must be the same D. And the reasoning is similar down the chain of intersections $B \between C \between B' \between ...$ so all the π-blocks and the σ-blocks in that chain of intersections must be contained in the same $D \in \tau$, thus that block of the meet $\pi \wedge \sigma$ (the union of those intersecting blocks) must be contained in that $D \in \tau$ so $\tau \precsim \pi \wedge \sigma$, i.e., the set-of-blocks defined meet is the greatest lower bound in the refinement partial ordering.

1.2.3   The graph-theoretic definitions of join and meet

Every partition π on U defines a graph $Gph(\pi)$ on the elements of U as the vertices (or nodes) and which is simple (at most one link or arc between any two vertices), undirected, and whose set of links or arcs is simply indit(π). Thus the graph $Gph(\mathbf{0}_U)$ associated with the indiscrete partition $\mathbf{0}_U$ is the complete graph $K(U)$ on U (a link between any two vertices plus a loop at each vertex) and $Gph(\mathbf{1}_U)$ is the graph with only the loops at the vertices.

The graph-theoretic approach to defining logical operations on partitions [30] starts with the complete graph $K(U)$. Then for any partition π on U and any link $u - u'$, mark the link with $T\pi$ if $(u, u') \in \text{dit}(\pi)$ and with $F\pi$ if $(u, u') \in \text{indit}(\pi)$. Doing the same with another partition σ on U will result in each link being labeled with 'truth-values' such as $u\xrightarrow{T\sigma, F\pi}u'$. Then for the join, meet, or any binary logical (truth-functional) operation, label the link with the appropriate truth value for that operation. For instance, for the join or disjunction, we have in Table 1.2:

π	σ	$\pi \vee \sigma$
$T\pi$	$T\sigma$	$T(\pi \vee \sigma)$
$T\pi$	$F\sigma$	$T(\pi \vee \sigma)$
$F\pi$	$T\sigma$	$T(\pi \vee \sigma)$
$F\pi$	$F\sigma$	$F(\pi \vee \sigma)$

Table 1.2: Truth table for the join.

Hence to define the partition join, we would label the link $u\xrightarrow{T\sigma, F\pi} u'$ with $T(\pi \vee \sigma)$ and so forth for all the links according to the truth table. Then we delete all the 'truth' links labeled with $T(\pi \vee \sigma)$ leaving only the 'false' links labeled with $F(\pi \vee \sigma)$. The connected components of that 'false-graph' $Gph(\pi \vee \sigma)$ is the partition on U defined by the truth-table for the join. For the meet, we use the truth-table for the meet or conjunction, and for any other binary or n-ary truth-functional operation, we use the truth-table for that operation to define the corresponding operation on partitions.

To see that this graph-theoretic approach gives the same operation for the join as the set-of-blocks or ditset definitions, we note that a link $u - u'$ is labeled $F(\pi \vee \sigma)$ if and only if it was labeled $F\pi$ and $F\sigma$ which means that $(u, u') \in \text{indit}(\pi) \cap \text{indit}(\sigma) = \text{indit}(\pi \vee \sigma)$ so the definitions are equivalent.

For the meet or conjunction operation, the truth table is Table 1.3:

π	σ	$\pi \wedge \sigma$
$T\pi$	$T\sigma$	$T(\pi \wedge \sigma)$
$T\pi$	$F\sigma$	$F(\pi \wedge \sigma)$
$F\pi$	$T\sigma$	$F(\pi \wedge \sigma)$
$F\pi$	$F\sigma$	$F(\pi \wedge \sigma)$

Table 1.3: Truth table for the meet.

so the links with an $F(\pi \wedge \sigma)$ assigned to them are the ones where (u, u') is an indit of π or σ or both, i.e., $(u, u') \in \text{indit}(\pi) \cup \text{indit}(\sigma)$. Now $\text{dit}(\pi \wedge \sigma) = \text{int}[\text{dit}(\pi) \cap \text{dit}(\sigma)]$ so $\text{indit}(\pi \wedge \sigma) = \overline{[\text{indit}(\pi) \cup \text{indit}(\sigma)]}$, i.e., $\text{indit}(\pi \wedge \sigma)$ is the smallest equivalence relation containing $\text{indit}(\pi) \cup \text{indit}(\sigma)$. But we have just shown that the $F_{\pi \wedge \sigma}$-links are the indits of $\text{indit}(\pi) \cup \text{indit}(\sigma)$. The partition determined by that graph is its connected components and those components are the same if we take the transitive closure of the graph (i.e., put in a link $u - u'$ if there is any path of links connecting u to u'), and that transitive closure just corresponds to the closure $\overline{[\text{indit}(\pi) \cup \text{indit}(\sigma)]} = \text{indit}(\pi \wedge \sigma)$. Hence the graph-theoretic definition of the partition meet agrees with the ditset definition.

For example, consider the two partitions $\pi = \{\{a, b\}, \{c, d\}\}$ and $\sigma = \{\{a\}, \{b, c, d\}\}$ on $U = \{a, b, c, d\}$. The complete graph with the labeled links prior to considering any particular logical operation is given in Figure 1.3.

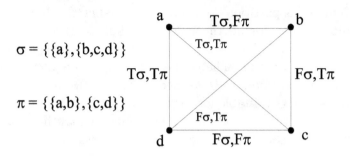

Figure 1.3: Graph with links labeled by truth-values from π and σ.

If we use the truth table for the join, than only the $d-c$ link has the $F\sigma, F\pi$ values for $F(\pi \vee \sigma)$ so it is the only surviving link when all the true-links are deleted and indicated by the thickened link in Figure 1.4. Then the connected components for $Gph(\pi \vee \sigma)$ are $\{a\}$, $\{b\}$, and $\{c,d\}$ which are the blocks of $\pi \vee \sigma$.

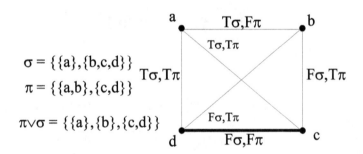

Figure 1.4: Connected components of graph with only $d - c$ link are blocks of $\pi \vee \sigma$.

If we use the truth table for the conjunction then the thickened links in Figure 1.5 are the false-links. The two links $a - d$ and $a - c$ are missing in the false-graph $Gph(\pi \wedge \sigma)$ but all the vertices are connected by false-links so there is only one connected component and thus the meet $\pi \wedge \sigma$ is the indiscrete partition $\mathbf{0}_U$.

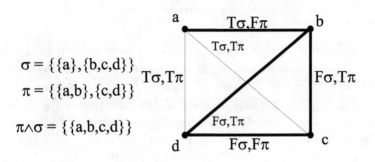

$$\sigma = \{\{a\},\{b,c,d\}\}$$

$$\pi = \{\{a,b\},\{c,d\}\}$$

$$\pi \wedge \sigma = \{\{a,b,c,d\}\}$$

Figure 1.5: All vertices connected in graph $Gph\,(\pi \wedge \sigma)$ so $\pi \wedge \sigma$ is the indiscrete partition.

1.2.4 The complete Boolean subalgebra definitions of join and meet

Oystein Ore [79] noted a one-to-one correspondence between complete Boolean subalgebras of the powerset Boolean algebra $\wp\,(U)$ and partitions on U. Given a partition π on U, the corresponding complete Boolean subalgebra $\mathcal{B}\,(\pi)$ is that generated by arbitrary unions of the blocks of π, and given a complete Boolean subalgebra $\mathcal{B}$ of $\wp\,(U)$, the atoms or minimal subsets form a partition $\pi_{\mathcal{B}}$ on U. By this approach, it is the meet that has the easiest definition:

$$\mathcal{B}\,(\pi \wedge \sigma) = \mathcal{B}\,(\pi) \cap \mathcal{B}\,(\sigma)$$

since the atoms or minimal subsets of that intersection Boolean algebra would be the smallest subsets that are a union of blocks of π and at the same time a union of blocks of σ. The union of two subalgebras is not necessarily a subalgebra but it generates one by taking arbitrary unions and intersections of subsets in the subalgebras. The atoms in that generated complete Boolean subalgebra are the non-empty intersections $B \cap C$ for $B \in \pi$ and $C \in \sigma$ so we have:

$$\mathcal{B}\,(\pi \vee \sigma) = \overline{(\mathcal{B}\,(\pi) \cup \mathcal{B}\,(\sigma))}.$$

1.2.5 The adjunctive characterizations of join and meet

The definitions can also be approached using the concepts of category-theoretic logic. Given two partial orders P and Q, two order-preserving functions $F : P \to Q$ and $G : Q \to P$ form an *adjunction* [5, p. 191] (or *Galois connection* [10, p. 124]) if for any $p \in P$ and $q \in Q$, the following equivalence holds:

$$F(p) \leq_Q q \text{ iff } p \leq_P G(q).$$

Then F is said to be the *left adjoint* and G the *right adjoint* in the adjunction. Taking $q = F(p)$, the element $GF(p) \in P$, called the *unit*, is the least element $G(q)$ such that $p \leq_P G(q)$. And taking $p = G(q)$, the element $FG(q) \in Q$, called the *counit*, is the greatest element $F(p)$ such that $F(p) \leq_Q q$.

For instance, take $P = \Pi(U)$ ordered by refinement and $Q = \wp(U \times U)$ ordered by inclusion with $F : P \to Q$ as $\pi \longmapsto \text{dit}(\pi)$ and $G : Q \to P$ as $V \longmapsto \lambda(\text{int}(V))$ where $\lambda(\text{int}(V))$ reads "the partition whose ditset is int (V)." Then we have the adjunction:

$$\text{dit}(\pi) \subseteq V \text{ iff } \pi \precsim \lambda(\text{int}(V)).$$

For $\pi \in \Pi(U)$, the unit is $\lambda(\text{int}(\text{dit}(\pi))) = \pi$ which is trivially the least $\lambda(\text{int}(V))$ such that $\pi \precsim \lambda(\text{int}(V))$. For $V \in \wp(U \times U)$, the counit is $\text{dit}(\lambda(\text{int}(V))) = \text{int}(V)$ which is the great ditset such that $\text{dit}(\pi) \subseteq V$. A subset $V \in \wp(U \times U)$ is said to be *open* if the over-and-back operation of taking V to the counit $\text{dit}(\lambda(\text{int}(V))) = \text{int}(V)$ is the identity, i.e., $V = \text{int}(V)$. The over-and-back operation of taking each π to its unit $\lambda(\text{int}(\text{dit}(\pi)))$ is already the identity so the maps F and G establish an isomorphism [10, p. 124, Theorem 20] between the partitions in $\Pi(U)$ and the open subsets in $\wp(U \times U)$ which we previously established:

$$\Pi(U) \cong \mathcal{O}(U \times U).$$

It might also be noted that the adjunction works just as well if we cut down the partial orders to the corresponding upper segments, $[\pi, \mathbf{1}_U]$ and $[\text{dit}(\pi), U \times U]$.

The powerset $\wp(U)$ is a category with the maps being the inclusion maps. There is then a diagonal functor $\Delta : \wp(U) \to \wp(U) \times \wp(U)$ from the powerset into the product category of the powerset times itself where $S \longmapsto (S, S)$. The diagonal functor has a left adjoint $\cup : \wp(U) \times \wp(U) \to \wp(U)$ that takes $(S, T) \longmapsto S \cup T$. That adjunction is equivalent to the statement:

$$S \cup T \subseteq W \text{ iff } (S, T) \subseteq (W, W).$$

Dually, the diagonal functor has a right adjoint that defines the intersection or meet: $\cap : \wp(U) \times \wp(U) \to \wp(U)$ where $(S, T) \longmapsto S \cap T$. That adjunction is equivalent to the statement:

$$(W, W) \subseteq (S, T) \text{ iff } W \subseteq S \cap T.$$

That is how the join and meet of sets can be defined using adjunctions [73, p. 96].

To apply this approach to partitions, we mimic it with the set of partitions $\Pi(U)$ on U partially ordered by refinement (or, equivalently, the set of ditsets $\mathcal{O}(U \times U)$ partially ordered by inclusion) replacing $\wp(U)$. Then the left adjoint to the diagonal functor $\tau \longmapsto (\tau, \tau)$ would satisfy:

$$\pi \vee \sigma \precsim \tau \text{ iff } (\pi, \sigma) \precsim (\tau, \tau)$$

which simply states that the join $\pi \vee \sigma$ is the least upper bound of π and σ in $\Pi(U)$ and that characterizes the join. Similarly, the right adjoint to the diagonal functor would satisfy:

$$(\tau, \tau) \precsim (\pi, \sigma) \text{ iff } \tau \precsim \pi \wedge \sigma$$

which simply states that the meet $\pi \wedge \sigma$ is the greatest lower bound of π and σ in $\Pi(U)$ and that characterizes the meet. The adjunctive approach thus characterizes the join and meet of partitions respectively as the least upper bound and greatest lower bound–which we have shown to exist by the previous methods.

1.3 The implication operation on partitions

1.3.1 Analogies with Heyting and bi-Heyting algebras

In the development of the logic of partitions, it is important to see analogies with a Heyting algebra (or intuitionistic propositional calculus) where the principal model is the algebra of open subsets of a topological space. We are treating the logic of partitions entirely from the semantic point of view, i.e., reasoning about partitions, not using a set of axioms. In that sense, the analogy is between the logic of open subsets of a topological space (the standard semantic model of a Heyting algebra) and the logic of partitions on a set. There are at best "analogies" since Heyting algebras are distributive lattices while the lattice of partitions is not distributive. Heyting algebras have a negation (i.e., the largest open subset disjoint from a given subset) and an implication (where the negation $\neg\sigma$ is the "implication to zero $\sigma \Rightarrow 0$"). The set of negated elements in a Heyting algebra (i.e., the regular open subsets in the standard topological semantic interpretation) form a Boolean algebra.

In the logic of partitions, there is also a negation and an implication (defined below) where the negation is the implication to zero $\sigma \Rightarrow \mathbf{0}_U$. In both cases, the negation can be relativized to a fixed element π (a given open subset or partition) where the π-negation is just the implication with a fixed consequent $\overset{\pi}{\neg}\sigma = \sigma \Rightarrow \pi$. In the logic of partitions as in the logic of open subsets, the negated and, in general, π-negated elements form a Boolean algebra.

In spite of the partition algebra not being distributive, there are some analogous results that are summarized in Table1.4 where $\overset{C}{\neg}B = B \Rightarrow C$ in a Heyting algebra.

Heyting Algebra	Partition Algebra
$B \leq C$ iff $(B \Rightarrow C) = 1$	$\sigma \precsim \pi$ iff $(\sigma \Rightarrow \pi) = 1_U$
$C \leq B \Rightarrow C$	$\pi \precsim \sigma \Rightarrow \pi$
$(1 \Rightarrow C) = C$	$(1_U \Rightarrow \pi) = \pi$
$(B \Rightarrow 1) = 1$	$(\sigma \Rightarrow 1_U) = 1_U$
$B \leq \overset{CC}{\neg\neg}B$	$\sigma \precsim \overset{\pi\pi}{\neg\neg}\sigma$
$\overset{C}{\neg}B = \overset{CCC}{\neg\neg\neg}B$	$\overset{\pi}{\neg}\sigma = \overset{\pi\pi\pi}{\neg\neg\neg}\sigma$
$\overset{C}{\neg}(B \vee \overset{C}{\neg}B) = 0$	$\overset{\pi}{\neg}(\sigma \vee \overset{\pi}{\neg}\sigma) = 0_U$
$\overset{C}{\neg}(B \vee C) = \overset{C}{\neg}B \wedge \overset{C}{\neg}C$	$\overset{\pi}{\neg}(\sigma \vee \tau) = \overset{\pi}{\neg}\sigma \wedge \overset{\pi}{\neg}\tau$

Table 1.4: Some analogous results in Heyting and Partition algebras

There is a formal axiomatic dual to the Heyting algebra axioms and it is the set of axioms for a co-Heyting algebra where the standard semantic model is the logic of closed subsets (complements of open subsets) of a topological space. The partitions on a set $\Pi(U)$ can be represented as the open subsets or ditsets of $U \times U$ when taken as a closure space with the RST-closure operation. The complements of the open subsets are the closed subsets or inditsets of $U \times U$ which are the equivalence relations on U. Hence there is a dual logic of equivalence relations analogous to the logic of closed subsets of a topological space. The dual to the implication operation on partitions or partition relations (ditsets) is the difference operation on equivalence relations. The negation is the "difference from 1" or $1 - \sigma$ and the relativized version is the "difference from π" or $\pi - \sigma$. That dual logic of equivalence relations will be outlined here but it contains nothing really new since it is just the logic of partitions looked at from a complementary point of view. Moreover, to compare formulas between the logic of subsets (all subsets in the Boolean case and open subsets in the intuitionistic case) and the logic of partitions, we need to stick to the approach initiated above (e.g., the logic of partition relations or ditsets), not the logic of equivalence relations. For instance, the standard formula for modus ponens, $(\sigma \wedge (\sigma \Rightarrow \pi)) \Rightarrow \pi$ is the unfamiliar formula $\pi - (\sigma \vee (\pi - \sigma))$ in the dual logic of equivalence relations. Hence our main development is along the lines of partitions represented by partition relations or ditsets–which is analogous to Heyting algebras or the logic of open subsets of a topological space.

Since co-Heyting algebras are axiomatized, one can consider other co-Heyting algebras that just the algebra of closed subsets ([67], [68]). Moreover, one can consider a bi-Heyting algebra or Heyting-Brouwer (HB) logic which has both implication and difference (or co-implication) operations ([87]; [114]). We will show that there is also a difference or co-implication operation on partitions–a dual structure on $\Pi(U)$–that is *not* just the complementary view in terms of equivalence relations. For instance, the operations on equivalence relations do not define any new operations on partitions since they are the same operations looked at in complementary terms. But the dual operations of co-negation and co-implication, based on the notion of modular partitions, do define new operations on partitions. The logic of partitions has some of the rich dual structure of a Heyting-Brouwer logic or bi-Heyting algebra except that it is more intricate and complex since partition logic is not distributive. Also, like the Boolean algebra of subsets, it is based on an arbitrary unstructured universe set U with no topologies or orderings.

1.3.2 The set-of-blocks definition of implication on partitions

In the Boolean algebra $\wp(U)$, for subsets $S, T \subseteq U$, the implication or conditional operation on subsets is defined as:

$$S \supset T = S^c \cup T.$$

One key property is that when the implication equals the top U of the Boolean algebra, then the partial order holds between the subsets:

$$(S \supset T) = U \text{ iff } S \subseteq T.$$

Another way to approach the set implication $S \supset T$ is to first consider indicator or characteristic function that indicates the degree to which S is *not* contained in T. That indicator function indicates 1 on $u \in U$ if u is an element of S that is not contained in T and indicates 0 otherwise. That function is just χ_{S-T}, the indicator function for $S - T = S \cap T^c$. Then the indicator function for the extent to which S *is* contained in T would be its negation:

$$1 - \chi_{S-T} = \chi_{(S-T)^c} = \chi_{S^c \cup T} = \chi_{S \supset T}.$$

Thus the set implication $S \supset T$ is a subset indicating the extent to which S is contained in T so when it is all true, i.e., $\chi_{S \supset T} = \chi_U$, then and only then $S \subseteq T$.

That means, in the partition case, that if the implication partition $\sigma \Rightarrow \pi$ was equal to the top, the discrete partition $\mathbf{1}_U$, then and only then would $\sigma \precsim \pi$ hold. That refinement relation holds if and only if for every $B \in \pi$, there is a $C \in \sigma$ such that $B \subseteq C$. One candidate definition that precisely satisfies that criterion is to take $\sigma \Rightarrow \pi$ to be the same as π except that when there is a $C \in \sigma$ such that $B \subseteq C$, then B is replaced by its singletons $\{u\}$ for $u \in B$. Thus if B is 'discretized' by being replaced by its singletons then it becomes the local version of the discrete partition $\mathbf{1}_B$. If B is not contained in any block of σ, then it remains as B which is the local version of the indiscrete partition $\mathbf{0}_B$. Hence in this definition of $\sigma \Rightarrow \pi$, the implication serves as a characteristic or indicator function with values or blocks $\mathbf{1}_B$ or $\mathbf{0}_B$ according to whether or not each block $B \in \pi$ is or is not contained in a block of σ. Hence this implication $\sigma \Rightarrow \pi$ indicates the degree to which π refines σ so when it is all true, then refinement holds:

$$\sigma \Rightarrow \pi = \mathbf{1}_U \text{ iff } \sigma \precsim \pi.$$

If $f, g : U \to \mathbb{R}$ are random variables on U, and $\pi = \left\{ f^{-1}(r) \right\}_{r \in f(U)}$ and $\sigma = \left\{ g^{-1}(r) \right\}_{r \in g(U)}$ are the inverse-image or coimage (or kernel) partitions determined by f and g, then when $\sigma \precsim \pi$, the random variable f is said to be a *sufficient statistic* for the random variable g, i.e., in an experiment, if the value of f is known, then that is sufficient to know the value of g [66, p, 31 where the opposite partial order is used.]. The implication $\sigma \Rightarrow \pi$ is thus a partition that indicates *the extent to which f is sufficient for g*, so when $\sigma \Rightarrow \pi = \mathbf{1}_U$, then f is sufficient for g.

No special attention need be paid to the complete Boolean subalgebra treatment of the partition implication since it is a variation of the set-of-blocks definition. The complete Boolean subalgebra $\mathcal{B}(\sigma \Rightarrow \pi)$ is generated from $\mathcal{B}(\pi)$ when each atom $B \in \pi$ is replaced by its discretization whenever there is a $C \in \sigma$ such that $B \subseteq C$.

1.3.3   The ditset definition of the partition implication

The ditset definition of the partition meet suggests a general way to define other logical operations on partitions: apply the set definition from subset logic to the ditsets (e.g., for the meet, take the intersection $\text{dit}(\pi) \cap \text{dit}(\sigma)$) and if the result is not a ditset, then take its interior as the ditset of the partition operation (e.g., $\text{dit}(\pi \wedge \sigma) = \text{int}[\text{dit}(\pi) \cap \text{dit}(\sigma)]$). Applying that method to the problem of defining the partition implication yields the definition:

$$\text{dit}(\sigma \Rightarrow \pi) = \text{int}[\text{dit}(\sigma)^c \cup \text{dit}(\pi)].$$

At first that ditset definition looks totally different from the set-of-blocks definition; it would threaten the naturalness of partition logic if there were several plausible definitions of the key operation of implication. But the definitions are the same. Let $\sigma \overset{*}{\Rightarrow} \pi$ temporarily stand for the set-of-blocks definition.

Proposition 1 $\sigma \Rightarrow \pi = \sigma \overset{*}{\Rightarrow} \pi$.

Proof: By the two definitions, $\text{dit}(\pi) \subseteq \text{dit}(\sigma \Rightarrow \pi) = \text{int}[\text{dit}(\sigma)^c \cup \text{dit}(\pi)]$ and $\text{dit}(\pi) \subseteq \text{dit}\left(\sigma \overset{*}{\Rightarrow} \pi\right)$ with the reverse inclusions holding between the inditsets. We prove the proposition by showing that $\text{dit}\left(\sigma \overset{*}{\Rightarrow} \pi\right) \subseteq \text{dit}(\sigma \Rightarrow \pi)$ and that $\text{indit}\left(\sigma \overset{*}{\Rightarrow} \pi\right) \subseteq \text{indit}(\sigma \Rightarrow \pi)$ where:

$$\text{indit}(\sigma \Rightarrow \pi) = \overline{\text{dit}(\sigma) \cap \text{dit}(\pi)^c} = \overline{(\text{indit}(\pi) \cap \text{indit}(\sigma)^c)} = \overline{(\text{indit}(\pi) - \text{indit}(\sigma))}.$$

Now suppose that $(u, u') \in \text{indit}\left(\sigma \overset{*}{\Rightarrow} \pi\right)$ (where $u \neq u'$) where $\text{indit}(\sigma * \Rightarrow \pi) \subseteq \text{indit}(\pi)$ so that $u, u' \in B$ for some block $B \in \pi$. Moreover if B were contained in any block $C \in \sigma$, then $(u, u') \in \text{dit}\left(\sigma \overset{*}{\Rightarrow} \pi\right) = \text{indit}\left(\sigma \overset{*}{\Rightarrow} \pi\right)^c$ contrary to assumption so B is not contained in any $C \in \sigma$. If u and u' were in different blocks of σ then $(u, u') \notin \text{indit}(\sigma)$ so that (u, u') would not be subtracted off in the formation of $\text{indit}(\sigma \Rightarrow \pi) = \overline{(\text{indit}(\pi) - \text{indit}(\sigma))}$ and thus would be in $\text{indit}(\sigma \Rightarrow \pi)$ which was to be shown. Now suppose that u and u' are in the same block $C \in \sigma$. Thus (u, u') was subtracted off in $\text{indit}(\pi) - \text{indit}(\sigma)$ and we need to show that it is restored in the closure $\overline{(\text{indit}(\pi) - \text{indit}(\sigma))}$. Since $u, u' \in B \cap C$ but B is not contained in any one block of σ, there is another σ-block C' such that $B \cap C' \neq \emptyset$. Let $u'' \in B \cap C'$. Then (u, u'') and (u', u'') are not in $\text{indit}(\sigma)$ since $u, u' \in C$ and $u'' \in C'$ but those two pairs are in $\text{indit}(\pi)$ since $u, u', u'' \in B$. Hence the pairs $(u, u''), (u', u'') \in \text{indit}(\pi) - \text{indit}(\sigma) = \text{indit}(\pi) \cap \text{dit}(\sigma)$ which implies that (u, u') must be in the closure $\text{indit}(\sigma \Rightarrow \pi) = \overline{(\text{indit}(\pi) - \text{indit}(\sigma))}$. That establishes $\text{indit}\left(\sigma \overset{*}{\Rightarrow} \pi\right) \subseteq \text{indit}(\sigma \Rightarrow \pi)$.

To prove the converse in the form $\text{dit}\left(\sigma \overset{*}{\Rightarrow} \pi\right) \subseteq \text{dit}(\sigma \Rightarrow \pi)$, assume $(u, u') \in \text{dit}\left(\sigma \overset{*}{\Rightarrow} \pi\right)$. Since $\text{dit}(\pi) \subseteq \text{dit}(\sigma \Rightarrow \pi)$, we would be finished if

$(u, u') \in \text{dit}(\pi)$. Hence assume $(u, u') \notin \text{dit}(\pi)$ so that $u, u' \in B$ for some π-block B and (u, u') is one of the new dits added when $\sigma \overset{*}{\Rightarrow} \pi$ is formed from π in the set-of-blocks definition. Thus $B \subseteq C$ for some σ-block C so that $(u, u') \in \text{indit}(\sigma)$ and (u, u') is not in the difference $\text{indit}(\pi) - \text{indit}(\sigma) = \text{indit}(\pi) \cap \text{dit}(\sigma)$. It remains to show that it is not in the closure $\text{indit}(\sigma \Rightarrow \pi) = \overline{(\text{indit}(\pi) - \text{indit}(\sigma))}$. To be in the closure, there would have to be some sequence $u = u_1, u_2, ..., u_n = u'$ such that $(u_i, u_{i+1}) \in \text{indit}(\pi) - \text{indit}(\sigma) = \text{indit}(\pi) \cap \text{dit}(\sigma)$ for $i = 1, ..., n-1$. But since all the $(u_i, u_{i+1}) \in \text{indit}(\pi)$ and $u = u_1 \in B$, all the $u = u_1, u_2, ..., u_n = u' \in B$ and $B \subseteq C$ so all the pairs $(u_i, u_{i+1}) \in \text{indit}(\sigma)$ which contradicts those pairs being in the difference $\text{indit}(\pi) - \text{indit}(\sigma) = \text{indit}(\pi) \cap \text{dit}(\sigma)$. Hence (u, u') is not in the closure $\text{indit}(\sigma \Rightarrow \pi) = \overline{(\text{indit}(\pi) - \text{indit}(\sigma))}$ so (u, u') is in the complement $\text{dit}(\sigma \Rightarrow \pi) = \text{indit}(\sigma \Rightarrow \pi)^c$ which completes the proof of the proposition. $\square$

Henceforth, $\sigma \Rightarrow \pi$ will refer to the partition implication defined either by the ditset definition or the set-of-blocks definition.

1.3.4 The graph-theoretic definition of the partition implication

To define the partition implication using the graph-theoretic method, we simply use the truth table for the implication (or conditional), Table 1.5, to label the links in $K(U)$ so that we may then eliminate the true-links to obtain $Gph(\sigma \Rightarrow \pi)$.

σ	π	$\sigma \Rightarrow \pi$
$T\sigma$	$T\pi$	$T(\sigma \Rightarrow \pi)$
$T\sigma$	$F\pi$	$F(\sigma \Rightarrow \pi)$
$F\sigma$	$T\pi$	$T(\sigma \Rightarrow \pi)$
$F\sigma$	$F\pi$	$T(\sigma \Rightarrow \pi)$

Table 1.5: Implication truth table for the implication.

Then the connected components of $Gph(\sigma \Rightarrow \pi)$ give the partition implication $\sigma \Rightarrow \pi$.

In the previous example of $\sigma = \{\{a\}, \{b, c, d\}\}$ and $\pi = \{\{a, b\}, \{c, d\}\}$, the only link of $K(U)$ labeled with $T\sigma$ and $F\pi$ was $a - b$ so it is the only $F(\sigma \Rightarrow \pi)$ to provide a link in $Gph(\sigma \Rightarrow \pi)$ and thus the connected components of $Gph(\sigma \Rightarrow \pi)$ are $\{a, b\}$, $\{c\}$, and $\{d\}$ so those are the blocks off $\sigma \Rightarrow \pi$ as in Figure 1.6.

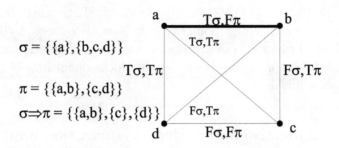

$\sigma = \{\{a\},\{b,c,d\}\}$

$\pi = \{\{a,b\},\{c,d\}\}$

$\sigma \Rightarrow \pi = \{\{a,b\},\{c\},\{d\}\}$

Figure 1.6: Graph-theoretic definition of $\sigma \Rightarrow \pi$.

Proposition 2 *Graph-theoretic definition = Set-of-blocks definition of $\sigma \Rightarrow \pi$.*

Proof: If $(u,u') \in$ dit (π), then T_π is assigned to that link in $K(U)$ so u and u' are not connected in $Gph(\sigma \Rightarrow \pi)$. And if $(u,u') \in$ indit (π) but also $(u,u') \in$ indit (σ), then $T(\sigma \Rightarrow \pi)$ is assigned to the link in $K(U)$ so again there is no connection between u and u' in $Gph(\sigma \Rightarrow \pi)$. There is a link $u - u'$ in $Gph(\sigma \Rightarrow \pi)$ in and only in the following situation where $(u,u') \in$ indit (π) and $(u,u') \in$ dit (σ)–which is exactly the situation when B is not contained in any block C of σ. Then for any other element $u'' \in B$ so that (u,u'') and $(u',u'') \in$ indit (π), u''has three possible locations relative to σ, in C, in C', or outside both as pictured in Figure 1.7.

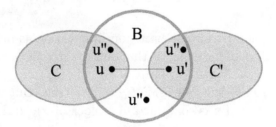

Figure 1.7: Possible locations for u''.

We must have either $(u,u'') \in$ dit (σ) or $(u',u'') \in$ dit (σ) so u'' is linked in $Gph(\sigma \Rightarrow \pi)$ to either u or to u', i.e., we have either $u\underline{{}^{T\sigma,F\pi}}u''$or $u'\underline{{}^{T\sigma,F\pi}}u''$. Thus all the elements of B are in the same connected component of the graph $Gph(\sigma \Rightarrow \pi)$ whenever B is not contained in any block of σ.

If, on the other hand, B is contained in some block C of σ, then any $u \in B$ cannot be linked to any other u' since that requires $F\pi$ assigned to the link $u - u'$ which requires $(u,u') \in$ indit (π), i.e., u,u' both in B. And since $B \subseteq C$, they both belong to C so $F\sigma$ and thus $T(\sigma \Rightarrow \pi)$ is also assigned to that link and hence that link is eliminated in $Gph\,(\sigma \Rightarrow \pi)$. Thus when B is contained in a block $C \in \sigma$, then any point $u \in B$ is a disconnected component to itself in $Gph\,(\sigma \Rightarrow \pi)$ so B is discretized in the graph-theoretic construction of $\sigma \Rightarrow \pi$. Thus the graph-theoretic and set-of-blocks definitions of the partition implication are equivalent. $\square$

1.3.5   The adjunctive approach to the partition implication

In the Boolean algebra of subsets, there is the "meet-with-S" functor $\wp\,(U) \to \wp\,(U)$ where $W \longmapsto W \cap S$ and the "implication-from-S" functor $\wp\,(U) \to \wp\,(U)$ where $T \longmapsto S \supset T$ which form an adjunction:

$$W \cap S \subseteq T \text{ iff } W \subseteq S \supset T$$

which characterizes $S \supset T$ as the largest subset such that modus ponens holds $((S \supset T) \cap S) \subseteq T$.

The analogous approach to the partition implication would be to find similar adjoints so that:

$$\tau \wedge \sigma \precsim \pi \text{ iff } \tau \precsim \sigma \Rightarrow \pi.$$

But this statement is false for the partition implication as defined above. The simplest non-trivial set of partitions is that on the three element set $U = \{a, b, c\}$ where we may take $\tau = \{\{a, b\}, \{c\}\}$, $\sigma = \{\{a, c\}, \{b\}\}$, and $\pi = \{\{a\}, \{b, c\}\}$. Then $\tau \wedge \sigma = \mathbf{0}_U$ so the left-hand side $\mathbf{0}_U \precsim \pi$ is true. But $\sigma \Rightarrow \pi = \pi$ (since no non-singleton block of π is contained in a block of σ), so the right-hand side is $\tau \precsim \pi$ which is false.

There is another way to see that the adjunction does not exist for partitions. In the Boolean algebra of subsets (or the Heyting algebra of open subsets of a topological space), that adjunction implies the distributivity of the algebra [14, p. 6]. But partition lattices $\Pi\,(U)$ are the standard examples of non-distributive lattices. In fact, it was an embarrassing moment for American mathematics in the nineteenth century when the mathematican-philosopher Charles Saunders Peirce claimed [82] to have proven the distributivity of all lattices but omitted the 'proof' as being too tedious. Europeans (e.g., Richard Dedekind and Ernest Schröder) soon besieged him with the example of the simplest non-trivial lattice on a three-element set shown in Figure 1.8.

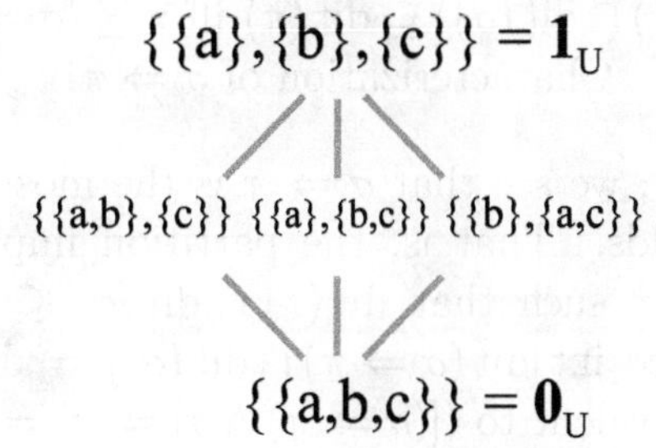

$$\{\{a\},\{b\},\{c\}\} = \mathbf{1}_U$$

$$\{\{a,b\},\{c\}\} \quad \{\{a\},\{b,c\}\} \quad \{\{b\},\{a,c\}\}$$

$$\{\{a,b,c\}\} = \mathbf{0}_U$$

Figure 1.8: Simplest non-distributive partition lattice on $U = \{a, b, c\}$.

Using the same τ, σ, and π, $\pi \vee \sigma = \mathbf{1}_U$ so $\tau \wedge (\pi \vee \sigma) = \tau$. But $\tau \wedge \pi = \tau \wedge \sigma = \mathbf{0}_U$ so $(\tau \wedge \pi) \vee (\tau \wedge \sigma) = \mathbf{0}_U$.

But there is another approach via adjunctions. Starting with the set $U \times U$, for any $S \subseteq U \times U$, there is the usual adjunction for the Boolean conditional $\wp(U \times U) \rightleftarrows \wp(U \times U)$:

$$(T \cap S) \subseteq P \text{ iff } T \subseteq (S \supset P)$$

for any subsets $T, P \in \wp(U \times U)$. Moreover, the dit-set representation $\Pi(U) \to \wp(U \times U)$ where $\tau \longmapsto \operatorname{dit}(\tau)$ has a right adjoint where $P \in \wp(U \times U)$ is taken to the partition $G(P)$ whose dit set is $\operatorname{int}(P)$:

$$\operatorname{dit}(\tau) \subseteq P \text{ iff } \tau \precsim G(P).$$

Composing the two right adjoints $\wp(U \times U) \to \wp(U \times U) \to \Pi(U)$ gives a functor taking $P \in \wp(U \times U)$ to $G_S(P)$ which is the partition whose dit set is $\operatorname{int}(S \supset P) = \operatorname{int}(S^c \cup P)$. Its left adjoint is obtained by composing the two left adjoints $\Pi(U) \to \wp(U \times U) \to \wp(U \times U)$ to obtain a functor taking a partition τ to $F_S(\tau) = \operatorname{dit}(\tau) \cap S$:

$$F_S(\tau) = (\operatorname{dit}(\tau) \cap S) \subseteq P \text{ iff } \operatorname{dit}(\tau) \subseteq (S \supset P) \text{ iff } \tau \precsim G_S(P).[5]$$

Specializing $S = \operatorname{dit}(\sigma)$ and $P = \operatorname{dit}(\pi)$ gives $G_{\operatorname{dit}(\sigma)}(\operatorname{dit}(\pi))$ as the partition whose ditset is $\operatorname{int}(\operatorname{dit}(\sigma)^c \cup \operatorname{dit}(\pi))$ which we know from above is the partition implication $\sigma \Rightarrow \pi$, i.e., $G_{\operatorname{dit}(\sigma)}(\operatorname{dit}(\pi)) = \sigma \Rightarrow \pi$. Using these restrictions, the adjunction gives the iff statement characterizing the partition implication:

---

[5]Thanks to Toby Kenney for suggesting this simplified presentation of this adjunction.

$$(\text{dit}\,(\tau) \cap \text{dit}\,(\sigma)) \subseteq \text{dit}\,(\pi) \text{ iff } \tau \preceq (\sigma \Rightarrow \pi).$$
$$\text{Characterization of } \sigma \Rightarrow \pi.$$

Taking $\tau = (\sigma \Rightarrow \pi)$, we see that $\sigma \Rightarrow \pi$ is the most refined partition such that modus ponens holds. That is, the partition implication $\sigma \Rightarrow \pi$ is the most refined partition τ such that $\text{dit}\,(\tau) \cap \text{dit}\,(\sigma) \subseteq \text{dit}\,(\pi)$ and thus that $((\sigma \Rightarrow \pi) \wedge \sigma) \precsim \pi$ since $\text{int}\,[\text{dit}\,(\sigma \Rightarrow \pi) \cap \text{dit}\,(\sigma)] = \text{dit}\,((\sigma \Rightarrow \pi) \wedge \sigma)$. And $(\sigma \Rightarrow \pi) \wedge \sigma \precsim \pi$ is equivalent to $([(\sigma \Rightarrow \pi) \wedge \sigma] \Rightarrow \pi) = \mathbf{1}_U$.

There are some other ways to state this characterization of the partition implication. Since arbitrary unions of ditsets are ditsets, we have:

$$\text{dit}\,(\sigma \Rightarrow \pi) = \bigcup \{\text{dit}\,(\tau) : \text{dit}\,(\tau) \cap \text{dit}\,(\sigma) \subseteq \text{dit}\,(\pi)\}.$$

Or taking the join in the complete lattice $\Pi\,(U)$,

$$\sigma \Rightarrow \pi = \bigvee \{\tau : \text{dit}\,(\tau) \cap \text{dit}\,(\sigma) \subseteq \text{dit}\,(\pi)\}.$$

Henceforth, the partition lattice $\Pi\,(U)$ equipped with the implication operation will be referred to as the *partition algebra* $\Pi\,(U)$.

1.4   Negation and other operations on partitions

1.4.1   Negation in partition logic

From the Boolean logic of subsets and the intuitionistic logic of open subsets, we have the suggestion to define the negation of a partition $\neg \sigma$ as the implication to zero, i.e., $\sigma \Rightarrow \mathbf{0}_U$. But in partition logic, this is immediately seen (using the set-of-blocks definition) to be rather trivial since for any $\sigma \neq \mathbf{0}_U$, $(\sigma \Rightarrow \mathbf{0}_U) = \mathbf{0}_U$ and for $\sigma = \mathbf{0}_U$, $(\mathbf{0}_U \Rightarrow \mathbf{0}_U) = \mathbf{1}_U$. In intuitionistic logic, the negation of an open subset is the largest open subset disjoint from the given open set. In the ditset representation, that would mean the ditset of the negation of a partition is the largest ditset disjoint from a given ditset. But in the partition case, for any $\sigma \neq \mathbf{0}_U$, i.e., for any non-empty ditset $\text{dit}\,(\sigma)$, there is *no* non-empty ditset disjoint from it. That means that any two non-empty ditsets always have a non-empty intersection, Thus for $\pi, \sigma \neq \mathbf{0}_U$, there is always a pair $\{u, u'\}$ that are in different blocks of *both* π and σ. This may also be stated as: if $\text{dit}\,(\pi) \cap \text{dit}\,(\sigma) = \emptyset$, then $\text{dit}\,(\pi) = \emptyset$ or $\text{dit}\,(\sigma) = \emptyset$. Or in terms of inditsets or equivalence relations: if $\text{indit}\,(\pi) \cup \text{indit}\,(\sigma) = U \times U$, then $\text{indit}\,(\pi) = U \times U$ or $\text{indit}\,(\sigma) = U \times U$.

That is an interesting result in its own right, and in view of the connections between partitions and graphs, it is a known result in graph theory. If $\sigma \neq \mathbf{0}_U$, then the graph on U with the links indit (σ) is disconnected. The largest graph on U with links disjoint from that indit (σ)-graph is the complementary graph whose links are given by dit(σ). The graph theorem is that the complementary graph of a disconnected graph is connected [113, p. 30], so with each vertex connected by links to every other vertex, the partition corresponding to a connected graph is $\mathbf{0}_U$, i.e., $\neg\sigma = \sigma \Rightarrow \mathbf{0}_U = \mathbf{0}_U$.

Theorem 1 (Common dits) *For any partitions $\pi \neq \mathbf{0}_U \neq \sigma$,*

$$\text{dit}\,(\pi) \cap \text{dit}\,(\pi) \neq \emptyset.$$

Proof: Since π is not the blob $\mathbf{0}_U$, consider two elements u and u' distinguished by π but identified by σ; otherwise $(u, u') \in \text{dit}\,(\pi) \cap \text{dit}\,(\sigma)$ and we are finished. Since σ is also not the blob, there must be a third element u'' not in the same block of σ as u and u'.

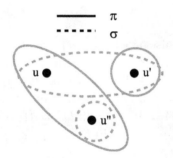

Figure 1.9: (u', u'') as a common dit

But since u and u' are in different blocks of π, the third element u'' must be distinguished from one or the other or both in π, e.g., distinguished from u' in both partitions as in Figure 1.9. Hence (u, u'') or (u', u'') must be distinguished by both partitions and thus must be in dit $(\pi) \cap$ dit (σ). $\qquad\square$

One implication is that if $\sigma \neq \mathbf{0}_U$, then $\overline{\text{dit}\,(\sigma)} = U \times U$, i.e., the RST-closure of any non-empty ditset is $U \times U$, since $\left(\overline{\text{dit}\,(\sigma)}\right)^c$ is the ditset of the largest ditset disjoint from dit (σ) but that is dit $(\mathbf{0}_U) = \emptyset$.

In spite of some similarities between partition logic and intuitionistic propositional logic, e.g., as modeled by open subsets of a topological space, there are some subtle differences. Both are lattices. In a lattice, a *pseudo-complement* [12, p.103] of an element σ (if it exists) is the maximal element σ^* such that $\sigma \wedge \sigma^* = 0$. If every element has a pseudo-complement, then the lattice is *pseudo-complemented*. In the lattice of open subsets of a topological space, the pseudo-complement of an open subset σ is the largest open subset disjoint from it, i.e., the interior of the complement, and that pseudo-complement might be denoted $\sigma \Rightarrow 0$. In the lattice of partitions $\Pi(U)$, for any partition σ, there is a largest ditset disjoint from dit (σ) and it is the ditset for the partition $\sigma \Rightarrow 0_U$. But this partition is not a pseudo-complement and $\Pi(U)$ is not pseudo-complemented. This is particularly striking since $\sigma \Rightarrow 0_U$ can be defined as having the largest ditset dit (τ) such that dit $(\sigma) \cap$ dit $(\tau) = \emptyset$. But that is *not* the same as the maximal partition τ such that $\sigma \wedge \tau = 0_U$ (recall that dit $(\sigma \wedge \tau) = \text{int}\,[\text{dit}\,(\sigma) \cap \text{dit}\,(\tau)]$).

Example: To show that $\sigma \Rightarrow 0_U$ is not the pseudo-complement of σ, let $\tau = \{\{a, b\}, \{c\}\}$ and $\sigma = \{\{a, c\}, \{b\}\}$ as in the previous example on $U = \{a, b, c\}$. Then $\tau \wedge \sigma = 0_U$ and $(\tau \wedge \sigma) \wedge ((\tau \wedge \sigma) \Rightarrow 0_U) = 0_U$. If $\sigma \Rightarrow 0_U$ was the pseudo-complement of σ, then it is the largest (most refined) partition π to satisfy $\pi \wedge \sigma = 0_U$ so

$$\tau \wedge ((\tau \wedge \sigma) \Rightarrow 0_U) \overset{?}{\precsim} \sigma \Rightarrow 0_U.$$

But that refinement is false since $(\tau \wedge \sigma) \Rightarrow 0_U = 0_U \Rightarrow 0_U = 1_U$ and $\tau \wedge 1_U = \tau$ so the LHS is $\tau \neq 0_U$ while the RHS is 0_U. The negation $\sigma \Rightarrow 0_U$ has the largest ditset disjoint from dit (σ) and $\sigma \wedge (\sigma \Rightarrow 0_U) = 0_U$. We also have $\tau \wedge ((\tau \wedge \sigma) \Rightarrow 0_U) \wedge \sigma = 0_U$, but the ditsets are not even disjoint since dit $(\tau \wedge ((\tau \wedge \sigma) \Rightarrow 0_U)) = \text{dit}\,(\tau) = \{(a, c), (b, c), ...\}$ and dit $(\sigma) = \{(a, b), (c, b), ...\}$ so   dit $(\tau \wedge ((\tau \wedge \sigma) \Rightarrow 0_U))$ $\cap$ dit (σ) $= \{(b, c), (c, b)\}$.

1.4.2 Relative negation in partition logic

In intuitionistic propositional logic (or in a Heyting algebra), the negated elements, called the *regular* elements (since they correspond to regular open sets, i.e., the interiors of closed sets, in the topological interpretation), form a Boolean algebra. In a partition algebra, the negated elements $\sigma \Rightarrow 0_U$ also form a Boolean algebra but is the trivial two-element one consisting of 0_U and 1_U. But the general implication $\sigma \Rightarrow \pi$ could be seen as an indicator function with values 1_B and 0_B indicating whether each block $B \in \pi$ is a subset of a block of σ or not. This suggests a non-trivial notion of negation where the indicator function's values are reversed, and that is given by the partition $(\sigma \Rightarrow \pi) \Rightarrow \pi$. Hence we consider the implication $\sigma \Rightarrow \pi$ as the *relative negation* of σ by π and denote this π-*negation* by $\overset{\pi}{\neg}\sigma = \sigma \Rightarrow \pi$. Then the formula $(\sigma \Rightarrow \pi) \Rightarrow \pi$ is just the double π-negation $\overset{\pi}{\neg}\overset{\pi}{\neg}\sigma$. The triple π-negation reverses the indicator values again to return to the original single π-negation so that:

$$\neg^{\pi}\neg^{\pi}\neg^{\pi}\sigma = \neg^{\pi}\sigma.$$

A π-negated partition is said to be π -*regular*. All the π-regular partitions are in the upper segment $[\pi, \mathbf{1}_U] = \{\tau : \pi \precsim \tau\}$, which is the lattice of partitions between π and $\mathbf{1}_U$ (including the endpoints). Let the set of π-regular partitions be denoted $\mathcal{B}[\pi, \mathbf{1}_U] \subseteq \Pi(U)$. We have just seen that the π-negation of a π-regular partition acts like the Boolean negation in $\mathcal{B}[\pi, \mathbf{1}_U]$ by considering each π-regular partition as an indicator function. Do the join and meet operations in $\Pi(U)$ act like the Boolean join and meet in $\mathcal{B}[\pi, \mathbf{1}_U]$? For any other partitions $\sigma, \tau \in \Pi(U)$, the partition join $\sigma \Rightarrow \pi \vee \tau \Rightarrow \pi$ of two π-regular partition can be analyzed using a truth table of its indicator values as in Table 1.6.

$\sigma \Rightarrow \pi$	$\tau \Rightarrow \pi$	$(\sigma \Rightarrow \pi) \vee (\tau \Rightarrow \pi)$
$\mathbf{1}_B$	$\mathbf{1}_B$	$\mathbf{1}_B$
$\mathbf{1}_B$	$\mathbf{0}_B$	$\mathbf{1}_B$
$\mathbf{0}_B$	$\mathbf{1}_B$	$\mathbf{1}_B$
$\mathbf{0}_B$	$\mathbf{0}_B$	$\mathbf{0}_B$

Table 1.6: Join operation on π-regular partitions

The blocks of the join are the non-empty intersections of the blocks of the two partitions. In any of the cases where a block B was discretized to $\mathbf{1}_B$, the singleton blocks $\{u\}$ for $u \in B$ would always yield the same singleton blocks when intersected with any other blocks so all the cases where a block B was atomized to $\mathbf{1}_B$, the join has that same value as indicated in Table 1.6 (the first three cases). If B was not atomized in either π-regular partition, then $B = \mathbf{0}_B$ would be the intersection in the join. But is $(\sigma \Rightarrow \pi) \vee (\tau \Rightarrow \pi)$ a π-regular partition in $\mathcal{B}[\pi, \mathbf{1}_U]$? Since $(\sigma \Rightarrow \pi) \vee (\tau \Rightarrow \pi)$ has values like a π-regular partition, it would equal its double π-negation which is a π-regular partition.

The same exercise can be carried out for the meet of two π-regular partitions.

$\sigma \Rightarrow \pi$	$\tau \Rightarrow \pi$	$(\sigma \Rightarrow \pi) \wedge (\tau \Rightarrow \pi)$
$\mathbf{1}_B$	$\mathbf{1}_B$	$\mathbf{1}_B$
$\mathbf{1}_B$	$\mathbf{0}_B$	$\mathbf{0}_B$
$\mathbf{0}_B$	$\mathbf{1}_B$	$\mathbf{0}_B$
$\mathbf{0}_B$	$\mathbf{0}_B$	$\mathbf{0}_B$

Table 1.7: Meet operation on π-regular partitions.

Whenever one of the values for a π-regular partition $\sigma \Rightarrow \pi$ or $\tau \Rightarrow \pi$ is $\mathbf{0}_B$ in the first two columns of Table 1.7, then it absorbs the other B-value $\mathbf{1}_B$ or $\mathbf{0}_B$, so the result for that row is also $\mathbf{0}_B$ (the last three rows). In the remaining case of two 1_B values, the singletons only intersect with themselves so that value is 1_B in the meet. Since the meet has only the values of 0_B and 1_B, its double π-negation has the same values and is a π-regular partition.

The exercise might also be carried out for the implication of two π-regular partitions as indicated in Table 1.8.

$\sigma \Rightarrow \pi$	$\tau \Rightarrow \pi$	$(\sigma \Rightarrow \pi) \Rightarrow (\tau \Rightarrow \pi)$
1_B	1_B	1_B
1_B	0_B	0_B
0_B	1_B	1_B
0_B	0_B	1_B

Table 1.8: Implication operation on π-regular partitions.

For any row where the consequent $\tau \Rightarrow \pi$ has the value 1_B, it is already atomized so that value for the overall implication is also 1_B. When both the values are 0_B (i.e., the last row), then since $B \subseteq B$, the overall implication has the value 1_B in that case. In the remaining case with antecedent 1_B and consequent 0_B, B is not contained in its atomized version, so it remains B and thus that row has the value 0_B. And it is equivalent to its double π-negation so that implication is also π-regular.

In this manner, we see that all the partition operations on the π-regular elements are the same as the Boolean operations on the indicator values 1_B and 0_B for $B \in \pi$, so $\mathcal{B}[\pi, 1_U]$ is a Boolean algebra under the partition operations and will be called the *Boolean core* for π. The Boolean core $\mathcal{B}[\pi, 1_U]$ is not technically a subalgebra since the Boolean core $\mathcal{B}[\pi, 1_U] \subseteq [\pi, 1_U] = \{\tau : \pi \precsim \tau\}$ is contained in the upper segment $[\pi, 1_U]$ of $\Pi(U)$ determined by π and the bottom of the Boolean core $\mathcal{B}[\pi, 1_U]$ is π. It is customary to rule out the degenerate Boolean algebra 1 where the top equals the bottom so that means $\pi \neq 1_U$ for the Boolean algebras $\mathcal{B}[\pi, 1_U]$.

If the partition π has any singleton blocks $B = \{u\}$ for $u \in U$, then they are always included in a block of any other partition. Moreover they are the same as their atomized version so $1_B = 0_B$ when B is a singleton. Thus the construction of the Boolean core $\mathcal{B}[\pi, 1_U]$ ignores singleton blocks; they are like a useless appendage onto any π-regular partition and do not change under negation. Let π_{ns} stand for π with the singleton blocks removed. Then the π-regular partitions $\varphi \in \mathcal{B}[\pi_{ns}, 1_U]$ are characterized by the characteristic or indicator function of their non-singleton blocks with values 1_B or 0_B, which could be viewed as an normal characteristic function $\chi(\varphi) : \pi_{ns} \to 2$. The Boolean core $\mathcal{B}[\pi_{ns}, 1_U]$ is isomorphic to the powerset Boolean algebra $\wp(\pi_{ns})$ under the correspondence: $\varphi \longleftrightarrow \chi(\varphi) : \pi_{ns} \to 2$.

We now have two Boolean algebras associated with each partition $\pi \in \Pi(U)$ so we should expect a close relationship. In the complete Boolean subalgebra approach to defining the lattice operations, $\mathcal{B}(\pi)$ was the Boolean subalgebra of $\wp(U)$ generated by the unions of the blocks of π, singleton or not, so it is isomorphic to the powerset Boolean algebra $\wp(\pi)$. The difference between $\mathcal{B}(\pi)$ and $\mathcal{B}[\pi_{ns}, 1_U]$ is solely due to the different treatment of the singletons since $0_B = 1_B$ for singleton blocks B. There are two versions of an arbitrary union of π-blocks in $\mathcal{B}(\pi)$, namely with or without a singleton block. Hence if we multiply $\mathcal{B}[\pi_{ns}, 1_U]$ by the two-element Boolean algebra 2 for every singleton block in π, then we obtain $\mathcal{B}(\pi)$:

$$\mathcal{B}(\pi) \cong \mathcal{B}[\pi_{ns}, 1_U] \times \prod_{\{u\} \in \pi} 2.$$

1.4.3 The Sheffer stroke, not-and, or nand operation on partitions

In addition to the lattice operations and implication, we will analyze the Sheffer stroke, not-and, or nand operation denoted $\pi | \sigma$. The ditset definition would be:

$$\text{dit}(\pi | \sigma) = \text{int}\left[(\text{dit}(\pi) \cap \text{dit}(\sigma))^c\right] = \text{int}\left[\text{indit}(\pi) \cup \text{indit}(\sigma)\right]$$

so that:

$$\text{indit}(\pi | \sigma) = \overline{[\text{dit}(\pi) \cap \text{dit}(\sigma)]}.$$

In the graph-theoretic definition, the links in $K(U)$ marked with $F(\pi | \sigma)$ would be those with $T\pi$ and $T\sigma$ as indicated in the truth-table Table 1.9 for the nand. After deleting all the links marked with $F\pi$ or $F\sigma$ to obtain $Gph(\pi | \sigma)$, the connected components would be the partition $\pi | \sigma$.

π	σ	$\pi | \sigma$
$T\pi$	$T\sigma$	$F(\pi | \sigma)$
$T\pi$	$F\sigma$	$T(\pi | \sigma)$
$F\pi$	$T\sigma$	$T(\pi | \sigma)$
$F\pi$	$F\sigma$	$T(\pi | \sigma)$

Table 1.9: Truth-table for the partition nand $\pi | \sigma$.

For the set-of-blocks definition, each $u \in U$ is contained in some block $B \cap C$ of $\pi \vee \sigma$ and each different u' is contained in a block $B' \cap C'$. If $B \neq B'$ and $C \neq C'$, then (u, u') is a dit of both π and σ so:

$(u, u') \in (B \cap C) \times (B' \cap C') \subseteq \text{dit}\,(\pi) \cap \text{dit}\,(\sigma) = (\text{indit}\,(\pi) \cup \text{indit}\,(\sigma))^c.$

Two vertices are connected in $Gph\,(\pi|\sigma)$ if and only if they are in the closure $\overline{\text{dit}\,(\pi) \cap \text{dit}\,(\sigma)}$ and thus they are a distinction if and only if they are in the complement of the closure which is the interior: $\text{int}\,[\text{indit}\,(\sigma) \cup \text{indit}\,(\tau)]$. Thus the graph-theoretic and ditset definitions of $\pi|\sigma$ agree. To obtain the set-of-blocks definition, note that when u and u' are linked in $Gph\,(\pi|\sigma)$ because $B \neq B'$ and $C \neq C'$, then all the elements of $B \cap C$ and $B' \cap C'$ are in the same block of the nand $\pi|\sigma$. But for any non-empty $B \cap C$, if there is *no* other block $B' \cap C'$ of the join with $B \neq B'$ and $C \neq C'$, then the elements of $B \cap C$ would not even be connected with each other so they would be singletons in the nand. Hence for the set-of-blocks definition of the nand $\pi|\sigma$, the blocks of the nand partition are formed by taking the unions of any join blocks $B \cap C$ and $B' \cap C'$ which differ in both "components" but by taking as singletons the elements of any $B \cap C$ which does not differ from any other join block in both components.

Example: Let $\pi = \{\{a, b\}, \{c, d, e\}\}$ and $\sigma = \{\{a, b, c\}, \{d, e\}\}$. In Figure 1.10, all the arcs in the complete graph $K\,(U)$ on five vertices are labeled according to the status of the two endpoints in the thickened lines. In the graph $Gph\,(\pi|\sigma)$ with only the thickened links, there are two connected components giving the blocks of the nand: $\pi|\sigma = \{\{a, b, d, e\}, \{c\}\}$.

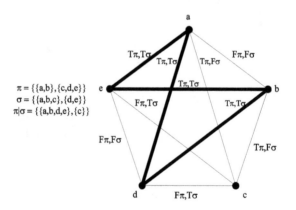

Figure 1.10: Graph-theoretic definition of $\pi|\sigma$.

By the set-of-blocks definition, the two blocks of the join $\{a, b\} = \{a, b\} \cap \{a, b, c\} = B \cap C$ and $\{d, e\} = \{c, d, e\} \cap \{d, e\} = B' \cap C'$ have both $B \neq B'$ and $C \neq C'$, so all those elements are in one block $\{a, b, d, e\}$ of the nand. And $\{c\} = \{c, d, e\} \cap \{a, b, c\}$ but there are no other blocks in the join $\pi \vee \sigma = \{\{a, b\}, \{c\}, \{d, e\}\}$ that differ in both blocks intersected, so $\{c\}$ is a singleton in the nand.

An *atom* in the lattice of partitions $\Pi(U)$ is a partition π so that there is no partition between it and the blob $\mathbf{0}_U$, i.e., if $\mathbf{0}_U \precsim \varphi \precsim \pi$, then $\varphi = \mathbf{0}_U$ or $\varphi = \pi$. The atoms are the partitions with exactly two blocks.

Example: An interesting special case are atoms where one block is a singleton such as $\pi = \{\{u\}, U - \{u\}\}$ and $\sigma = \{\{u'\}, U - \{u'\}\}$ for $u \neq u'$. Then the join has only three blocks $\pi \vee \sigma = \{\{u\}, \{u'\}, U - \{u, u'\}\}$ where $\{u\} = \{u\} \cap (U - \{u'\})$ and $\{u'\} = (U - \{u\}) \cap \{u'\}$. Since those intersections differ in both components, $\{u, u'\}$ is a block in the nand. But $U - \{u, u'\} = U - \{u\} \cap U - \{u'\}$ and there are no other non-empty intersections that differ in both blocks intersected so that block $U - \{u, u'\}$ is atomized or discretized in the nand, i.e., $\pi | \sigma = \{\{u, u'\}, \{u''\}, ...\}$. On the graph-theoretic approach, the only link $F_{\pi|\sigma}$ in $Gph(\pi|\sigma)$ is $u \underline{\quad T\pi, T\sigma \quad} u'$ so u and u' are in the same block and all other $u'' \in U$ are singletons. Thus $\pi | \sigma$ is a coatom of $\Pi(U)$, i.e., a partition so there is no partition strictly between it and the top $\mathbf{1}_U$.

Example: The universe set $U = \{Tom, John, Jim\}$ consists of three people and there are two partitions: α which distinguishes people according to the first letter of their name so that $\alpha = \{\{Tom\}, \{John, Jim\}\}$, and ω which distinguishes people according to the last letter of their name so that $\omega = \{\{Tom, Jim\}, \{John\}\}$. Then the meet $\alpha \wedge \omega$ would identify people who are directly and indirectly identified by the two partitions. Tom and John are not directly identified but are indirectly identified: $Tom \overset{\omega}{\sim} Jim \overset{\alpha}{\sim} John$ so that $\sigma \wedge \omega = \mathbf{0}_U$. But since the meet is $\mathbf{0}_U$, the orthogonal nand of the two partitions could be non-zero, and in fact $\alpha | \omega = \{\{Tom, John\}, \{Jim\}\}$. Thus the fact that Tom and John are directly distinguished by both the first and last letters of their names, i.e., $T_\alpha T_\omega$ on the link $Tom - John$, results in them not being distinguished, i.e., $F_{\alpha|\omega}$. by the nand partition. In this case, $\neg\alpha \vee \neg\omega = \mathbf{0}_U$ and $\neg(\alpha \wedge \omega) = \mathbf{1}_U$.

A number of results about the nand can be obtained using the graph-theoretic characterization. Truth means distinctions. Hence for any ordered pair $(u, u') \in \text{dit}(\pi) \cap \text{dit}(\sigma)$, the $u - u'$ link in $K(U)$ would be marked $T\pi, T\sigma$ and thus $F_{\pi|\sigma}$ so u and u' will be in the same block of $\pi|\sigma$. By the Common-Dits result, any two non-blob partitions have dits in common, so for any $\pi, \sigma \neq \mathbf{0}_U$, $\pi|\sigma \neq \mathbf{1}_U$. Moreover, if $\pi|\sigma \neq \mathbf{1}_U$, then for some link in $Gph(\pi|\sigma)$ has $T\pi T\sigma$ assigned to it so that the Common-Dits Theorem implies $\pi, \sigma \neq \mathbf{0}_U$. And if π or σ is $\mathbf{0}_U$, then there can be no common dits (since the blob has no distinctions), so for any π:

$$\pi | \mathbf{0}_U = \mathbf{1}_U.$$

That is the partition version of the set relation $S|\emptyset = U$ since for $S, T \subseteq U$,

$S|T = S^c \cup T^c$.

In subset logic, negation can be defined in terms of the nand: $S|S = S^c = S \supset \emptyset$. In partition logic, for $\sigma = \mathbf{0}_U$, we have already noted that $\mathbf{0}_U|\mathbf{0}_U = \mathbf{1}_U = \mathbf{0}_U \Rightarrow \mathbf{0}_U$. If $\sigma \neq \mathbf{0}_U$, then for any two blocks $C, C' \in \sigma$ with $u \in C$ and $u' \in C'$, then the link $u - u'$ has T_σ, T_σ assigned to it and thus $F_{\sigma|\sigma}$ so that u and u' are in the same block of $\sigma|\sigma$. Similarly any other $u'' \in C$ is linked to u' so u and u'' are in the same connected component of $Gph(\sigma|\sigma)$. By symmetry, any other $u''' \in C'$ is in the same connected component, and this is true for any $C, C' \in \sigma$, so we have:

$$\sigma|\sigma = \mathbf{0}_U = \sigma \Rightarrow \mathbf{0}_U = \neg\sigma.$$

In subset logic, two subsets are disjoint (or orthogonal) if their meet (intersection) is the bottom zero element $\emptyset$, or equivalently (by the DeMorgan laws) if the union of their complements was the top element U. But in partition logic, these relationship are more subtle. Two partitions φ and φ' on U are said to be π-orthogonal if $\overset{\pi}{\neg}\varphi \vee \overset{\pi}{\neg}\varphi' = \mathbf{1}_U$. They are orthogonal if $\neg\varphi \vee \neg\varphi' = \mathbf{1}_U$. Orthogonality and π-orthogonality give a partition version of "disjointness."

Lemma 2 φ and φ' are orthogonal, i.e., $\neg\varphi \vee \neg\varphi' = \mathbf{1}_U$, iff $\varphi|\varphi' = \mathbf{1}_U$.

Proof: If $\neg\varphi \vee \neg\varphi' = \mathbf{1}_U$, then $int(indit(\varphi)) \cup int(indit(\varphi')) = dit(\mathbf{1}_U) = U^2 - \Delta$. By the monotonicity of the interior operator,

$$int(indit(\varphi)) \cup int(indit(\varphi')) \subseteq int(indit(\varphi) \cup indit(\varphi')) = dit(\varphi|\varphi')$$

so $\varphi|\varphi' = \mathbf{1}_U$. Conversely, if $\varphi|\varphi' = \mathbf{1}_U$, then one of the partitions have to be the blob $\mathbf{0}_U$ so $\neg\varphi$ or $\neg\varphi'$ is $\mathbf{1}_U$ and so is the join. $\square$

The contrapositive (negation of each side) of the Lemma is just a restatement of the Common-dits Theorem since $\neg\varphi \vee \neg\varphi' \neq \mathbf{1}_U$ just says that neither φ nor φ' is the blob $\mathbf{0}_U$ and the other side of the equivalence $\varphi|\varphi' \neq \mathbf{1}_U$ just says that $dit(\varphi) \cap dit(\varphi') \neq \emptyset$.

Thus orthogonality is also characterized by: $\varphi|\varphi' = \mathbf{1}_U$. And thus orthogonality means that if one partition is not $\mathbf{0}_U$, then the other must be $\mathbf{0}_U$. Orthogonality thus immediately implies $\varphi \wedge \varphi' = \mathbf{0}_U$ but not the reverse. In the previous example, the meet of $\pi = \{\{u\}, U - \{u\}\}$ and $\sigma = \{\{u'\}, U - \{u'\}\}$ is $\pi \wedge \sigma = \mathbf{0}_U$ and $\neg\mathbf{0}_U = \mathbf{1}_U$ but $\pi|\sigma \neq \mathbf{1}_U$ ($\{u, u'\}$ is a block in the nand) since neither is the blob. Also this means that the negation $\neg(\pi \wedge \sigma)$ is not necessarily the same as the nand $\pi|\sigma$. Since $dit(\pi \wedge \sigma) = int[dit(\pi) \cap dit(\sigma)]$ and $dit(\pi|\sigma) = int[(dit(\pi) \cap dit(\sigma))^c]$, they cannot have a dit in common, so their nand must be the discrete partition:

$$(\pi \wedge \sigma) \,|\, (\pi|\sigma) = \mathbf{1}_U$$
"and" orthogonal to (or "disjoint" with) "nand".

The same example shows that $\neg\pi \vee \neg\sigma$ (which is $\mathbf{0}_U$) is not necessarily the same as $\pi|\sigma$. The formulas are equivalent in subset logic, but in partition logic, the relationship is:

$$\neg\pi \vee \neg\sigma \precsim \pi|\sigma \precsim \neg(\pi \wedge \sigma)$$

since

$$\text{int}\,[\text{indit}\,(\pi)] \cup \text{int}\,[\text{indit}\,(\sigma)] \subseteq \text{int}\,[\text{indit}\,(\pi) \cup \text{indit}\,(\sigma)]$$
$$\subseteq \text{int}\,[\text{indit}\,(\pi \wedge \sigma)] = \text{int}\left[\overline{\text{indit}\,(\pi) \cup \text{indit}\,(\sigma)}\right].$$

Hence the "strong" DeMorgan law $\neg\sigma \vee \neg\tau = \neg(\sigma \wedge \tau)$ does not hold in partition logic. But the weak DeMorgan law, $\neg(\pi \vee \sigma) = \neg\pi \wedge \neg\sigma$, does hold (since if $\pi = \mathbf{0}_U = \sigma$, then both sides are $\mathbf{1}_U$, and if one is not $\mathbf{0}_U$, then both sides are $\mathbf{0}_U$).

Since the three formulas $\neg\pi \vee \neg\sigma$, $\pi|\sigma$, and $\neg(\pi \wedge \sigma)$ are classically equivalent, i.e., in subset logic, they will have the same truth table, and the truth table is used to define the *atomic* partition operations in the graph-theoretic method. That does not mean that the three formulas are equivalent in partition logic since the two formulas,$\neg\pi \vee \neg\sigma$ and $\neg(\pi \wedge \sigma)$, are not atomic but are compound formulas using other atomic partition operations. In subset logic, there are 16 different truth tables for binary operations, and all compound formulas with only two variables will have a truth table equivalent to one of the 16 possible atomic operations. In fact, there are smaller subsets of atomic operations that can be used to define all the others in subset logic, the most famous being the nand operation (or its negation neither-nor) by itself. But we have just seen that the binary operations on partitions are not closed under the 16 atomic definitions definable by the graph-theoretic method from the 16 possible truth tables. For instance, $\neg\pi \vee \neg\sigma$ and $\neg(\pi \wedge \sigma)$ also define new binary operations on partitions–but not atomic ones.

Partitions are more complex objects than subsets, and, accordingly, partition logic is much more complex than subset logic. We will see below that four atomic partition operations suffice to define all 16 of the truth-table definable partition atomic operations, but all the compound formulas with only two variables define a much larger universe of binary partition operations. Since there is a countably infinite number of finite compound binary formulas, it is not even currently known if the number of binary partition operations defined by the 16 logical atomic operations is infinite or finite.

In contrast to binary partition logical operations, the unary (including the 0-ary constants) operations are quite manageable. There are four possible truth tables for unary operations, which give $\mathbf{0}_U, \sigma, \neg\sigma$, and $\mathbf{1}_U$. These are the four graph-theoretic definable atomic unary operations. What happens when we form compound formulas? The negation applied to the top or bottom just gives the other, and the negation of σ is $\neg\sigma$. But the double negation $\neg\neg\sigma$ is not necessarily the same as σ so it is a new unary operation. But the triple negation is the same as the single negation so the number of unary operations, atomic or compound are the five ones, the four atomic ones $\mathbf{0}_U, \sigma, \neg\sigma$, and $\mathbf{1}_U$ and the compound operation $\neg\neg\sigma$. If we allow other binary operations, then there are other unary operations such as: $\sigma \vee \neg\sigma$.

Just as the unary operation $\neg\sigma$ is usefully generalized by the binary operation $\overset{\pi}{\neg}\sigma = \sigma \Rightarrow \pi$, so the binary operation $\sigma|\tau$ might be usefully generalized by the ternary operation $\sigma|_\pi\tau$, which is the nand operation relative to π. The basic idea is that the nand relativized by π is like the usual nand but relativized to the upper segment $[\pi, \mathbf{1}_U]$ defined by π. Hence its ditset must include dit (π) so the natural change from dit $(\sigma|\tau) = \text{int} [\text{indit} (\sigma) \cup \text{indit} (\tau)]$ to the ditset definition of $\sigma|_\pi\tau$ is:

$$\text{dit} (\sigma|_\pi\tau) = \text{int} [\text{indit} (\sigma) \cup \text{indit} (\tau) \cup \text{dit} (\pi)].$$

Since all the ditset definitions can be seen as being the interior of some Boolean subset combination of ditsets and inditsets (whether the interior operation is needed or not), the inditsets are always the RST-closures of negated Boolean combination of inditsets and ditsets, e.g.,

$$\text{indit} (\sigma|_\pi\tau) = \overline{[\text{dit} (\sigma) \cap \text{dit} (\tau) \cap \text{indit} (\pi)]}.$$

That closure operation corresponds to taking the transitive closure of the connected component in the graph $Gph (\sigma|_\pi\tau)$ for the equivalent graph-theoretic definition of the ternary operation. Hence we can read off from the ditset definition's inditset what the $F_{\sigma|_\pi\tau}$ conditions are, namely just $T\sigma$, $T\tau$, and $F\pi$. Hence we see that the truth-table for that ternary operation is given in Table 1.10 which is the truth table for the classical (not partition) formula $(\sigma|\tau) \vee \pi$.

σ	τ	π	$(\sigma\|\tau)\vee\pi$
T	T	T	T
T	T	F	F
T	F	T	T
T	F	F	T
F	T	T	T
F	T	F	T
F	F	T	T
F	F	F	T

Table 1.10: Truth table for graph-theoretic definition of $\sigma|_\pi\tau$.

Starting with that truth table, the graph $Gph\,(\sigma|_\pi\tau)$ would only have $F_{\sigma|_\pi\tau}$ links with $T\sigma T\tau F\pi$ on them so the transitive closure would be

$$\mathrm{indit}\,(\sigma|_\pi\tau) = \overline{[\mathrm{dit}\,(\sigma)\cap\mathrm{dit}\,(\tau)\cap\mathrm{indit}\,(\pi)]}$$

and thus the ditset and graph-theoretic definitions agree.

Since $\mathrm{dit}\,(\pi)$ was included in the ditset definition,

$$\mathrm{dit}\,(\sigma|_\pi\tau) = \mathrm{int}\,[\mathrm{indit}\,(\sigma)\cup\mathrm{indit}\,(\tau)\cup\mathrm{dit}\,(\pi)],$$

we have $\pi\precsim\sigma|_\pi\tau$. As in the case of the unrelativized nand, we have three classically equivalent formulas, $(\sigma\Rightarrow\pi)\vee(\tau\Rightarrow\pi) = \overset{\pi}{\neg}\sigma\vee\overset{\pi}{\neg}\tau$, $\sigma|_\pi\tau$, and $(\sigma\wedge\tau)\Rightarrow\pi = \overset{\pi}{\neg}(\sigma\wedge\tau)$ that are different in partition logic where:

$$\overset{\pi}{\neg}\sigma\vee\overset{\pi}{\neg}\tau\precsim\sigma|_\pi\tau\precsim\overset{\pi}{\neg}(\sigma\wedge\tau)$$

since:

$$\mathrm{int}\,(\mathrm{indit}\,(\sigma)\cup\mathrm{dit}\,(\pi))\cup\mathrm{int}\,(\mathrm{indit}\,(\tau)\cup\mathrm{dit}\,(\pi))$$
$$\subseteq\mathrm{int}\,[\mathrm{indit}\,(\sigma)\cup\mathrm{indit}\,(\tau)\cup\mathrm{dit}\,(\pi)]$$
$$\subseteq\mathrm{int}\,[\mathrm{int}\,(\mathrm{dit}\,(\sigma)\cap\mathrm{dit}\,(\tau))^c\cup\mathrm{dit}\,(\pi)]$$
$$=\mathrm{int}\,\left[\overline{\mathrm{indit}\,(\sigma)\cup\mathrm{indit}\,(\tau)}\cup\mathrm{dit}\,(\pi)\right].$$

It is immediate that

$$\mathrm{dit}\,(\sigma|_\pi\sigma) = \mathrm{int}\,[\mathrm{int}\,(\mathrm{dit}\,(\sigma)\cap\mathrm{dit}\,(\sigma))^c\cup\mathrm{dit}\,(\pi)]$$
$$=\mathrm{int}\,[\mathrm{indit}\,(\sigma)\cup\mathrm{dit}\,(\pi)] = \mathrm{dit}\,(\sigma\Rightarrow\pi)$$

so that:

$$\sigma|_\pi \sigma = \overset{\pi}{\neg}\sigma.$$

It is also clear that:

$$\sigma|_\pi \pi = \mathbf{1}_U$$

since dit $(\sigma|_\pi\pi) = \text{int}\left[\text{indit}(\sigma) \cup \text{indit}(\pi) \cup \text{dit}(\pi)\right] = \text{int}\left[U \times U\right] = \text{dit}(\mathbf{1}_U)$.

As previously defined, two partitions σ and τ on U are said to be π-orthogonal if $\overset{\pi}{\neg}\sigma \vee \overset{\pi}{\neg}\tau = \mathbf{1}_U$. The double π-negation $\overset{\pi}{\neg}\overset{\pi}{\neg}\sigma$ of any partition σ is its Booleanization, i.e., the closest element of the Boolean core $\mathcal{B}[\pi, \mathbf{1}_U]$ to σ in the sense that $\sigma \precsim \overset{\pi}{\neg}\overset{\pi}{\neg}\sigma$ and for any other $\overset{\pi}{\neg}\varphi \in \mathcal{B}[\pi, \mathbf{1}_U]$ with $\sigma \precsim \overset{\pi}{\neg}\varphi$, then $\overset{\pi}{\neg}\overset{\pi}{\neg}\sigma \precsim \overset{\pi}{\neg}\varphi$. This Booleanization determines a characteristic function $\chi_\sigma : \pi \to 2$ where $\chi_\sigma(B) = 0$ if the B-component of $\overset{\pi}{\neg}\overset{\pi}{\neg}\sigma$ was $\mathbf{0}_B$ (i.e., B is contained in a block of σ) and $\chi_\sigma(B) = 1$ if the B-component of $\overset{\pi}{\neg}\overset{\pi}{\neg}\sigma$ was $\mathbf{1}_B$ (i.e., the discretized version of B).

Lemma 3 σ and τ are π-orthogonal iff $\sum_{B\in\pi} \chi_\sigma(B) \chi_\tau(B) = 0$.

Proof: σ and τ are π-orthogonal iff each $B \in \pi$ is contained in block of σ or a block of τ iff for each $B \in \pi$, $\chi_\sigma(B) \chi_\tau(B) = 0$. $\square$

Every partition σ and its π-negation $\overset{\pi}{\neg}\sigma$ are π-orthogonal since $\overset{\pi}{\neg}\sigma \vee \overset{\pi}{\neg}\overset{\pi}{\neg}\sigma = \mathbf{1}_U$ which is just the law of excluded middle in the Boolean algebra $\mathcal{B}[\pi, \mathbf{1}_U]$.

Lemma 4 σ and τ are π-orthogonal, i.e., $\overset{\pi}{\neg}\sigma \vee \overset{\pi}{\neg}\tau = \mathbf{1}_U$, iff $\sigma|_\pi\tau = \mathbf{1}_U$.

Proof: Now $\overset{\pi}{\neg}\sigma \vee \overset{\pi}{\neg}\tau = \mathbf{1}_U$ iff for every $B \in \pi$, there is a $C \in \sigma$ or a $D \in \tau$, such that $B \subseteq C$ or $B \subseteq D$. Then for any link $u - u'$ with F_π assigned to it, i.e., $u, u' \in B$, $(u, u') \notin \text{dit}(\sigma) \cap \text{dit}(\tau)$, so the link $u - u'$ could not have $F_{\sigma|_\pi\tau}$ assigned to it in $Gph(\sigma|_\pi\tau)$ so $\sigma|_\pi\tau = \mathbf{1}_U$. Conversely if $\sigma|_\pi\tau = \mathbf{1}_U$, then for any two $u, u' \in B \in \pi$, we much have F_σ or F_τ assigned to the link $u - u'$ which means that $u, u \in C$ for some $C \in \sigma$ or $u, u' \in D$ for some $D \in \tau$. But u, u' were arbitrary $u, u' \in B$ so all the elements of B must belong to the same C or the same D, and that holds for any $B \in \pi$, so $\overset{\pi}{\neg}\sigma \vee \overset{\pi}{\neg}\tau = \mathbf{1}_U$. $\square$

The contrapositive (negation of both sides) of the Lemma is worth proving directly as a result of the Common Dits Theorem.

Corollary 1 $\overset{\pi}{\neg}\sigma \vee \overset{\pi}{\neg}\tau \neq \mathbf{1}_U$ iff $\sigma|_\pi\tau \neq \mathbf{1}_U$.

Proof: If $\overset{\pi}{\neg}\sigma \vee \overset{\pi}{\neg}\tau \neq \mathbf{1}_U$, then there is a $B \in \pi$ that not contained in any $C \in \sigma$ or any $D \in \tau$ (which implies B is not a singleton) so the partition σ restricted to B, denoted $\sigma \restriction B = \{B \cap C \neq \emptyset : C \in \sigma\}$ and similarly the partition τ restricted to B, $\tau \restriction B$ are neither the blob when the universe is B. Hence the Common-dits Theorem applied to the universe B with the two non-blob partitions on it implies that $\operatorname{dit}(\sigma \restriction B) \cap \operatorname{dit}(\tau \restriction B) \neq \emptyset$ so such a distinction (u, u') is in $\operatorname{dit}(\sigma) \cap \operatorname{dit}(\tau) \cap \operatorname{indit}(\pi)$ so $\sigma|_\pi \tau \neq 1_U$. Conversely, $\sigma|_\pi \tau \neq 1_U$ implies such a pair $(u, u') \in \operatorname{dit}(\sigma) \cap \operatorname{dit}(\tau) \cap \operatorname{indit}(\pi)$ and $(u, u') \in \operatorname{indit}(\pi)$ means there is a $B \in \pi$ such that $u, u' \in B$ while $(u, u') \in \operatorname{dit}(\sigma) \cap \operatorname{dit}(\tau)$. Hence we could not have $B \subseteq C$ (since then $B \cap C = B$) for any $C \in \sigma$ nor have $B \subseteq D$ for any $D \in \tau$, and thus $\overset{\pi}{\neg}\sigma \vee \overset{\pi}{\neg}\tau \neq 1_U$. $\qquad\square$

Note that while $\overset{\pi}{\neg}\sigma \vee \overset{\pi}{\neg}\tau \precsim \sigma|_\pi \tau \precsim \overset{\pi}{\neg}(\sigma \wedge \tau)$ and both $\overset{\pi}{\neg}\sigma \vee \overset{\pi}{\neg}\tau$ and $\overset{\pi}{\neg}(\sigma \wedge \tau)$ are in $\mathcal{B}[\pi, 1_U]$, there is no necessity for $\sigma|_\pi \tau$ to be in $\mathcal{B}[\pi, 1_U]$ as illustrated in Figure 1.11 even though it is always in the upper segment $[\pi, 1_U]$.

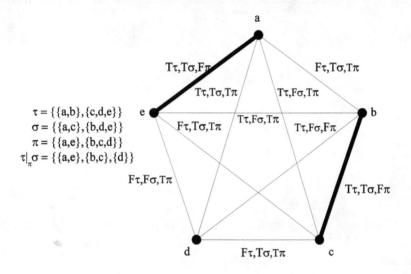

Figure 1.11: Example where $\tau|_\pi \sigma$ not in $\mathcal{B}[\pi, 1_U]$.

In the example of Figure 1.11, $\overset{\pi}{\neg}\tau \vee \overset{\pi}{\neg}\sigma = \pi$ and $\overset{\pi}{\neg}(\tau \wedge \sigma) = 1_U$, but $\pi \precsim \tau|_\pi \sigma = \{\{a, e\}, \{b, c\}, \{d\}\} \notin \mathcal{B}[\pi, 1_U]$.

1.4.4 The sixteen binary logical operations on partitions

There are $2^{\left(2^2\right)} = 16$ truth tables for binary operations. In terms of subsets $S, T \subseteq U$, the four rows in the truth table for a binary operation correspond to: $S \cap T$, $S \cap T^c$, $S^c \cap T$, and $S^c \cap T^c$ which give the four atomic areas in the Venn diagram for S and T. Then there are $2^4 = 16$ possible unions of some of the atomic areas to give the 16 logically definable subsets in terms of S and T.

Taking the universe to be $U \times U$ and S, T to be dit (π) and dit (σ), there are again 16 subsets definable in terms of these ditsets and their complementary inditsets. Some of those subsets will already be a ditset, e.g., dit $(\pi) \cup$ dit $(\sigma) =$ dit $(\pi \vee \sigma)$, but most will require the interior operation to obtain a ditset, e.g., int [dit $(\pi) \cap$ dit (σ)] $=$ dit $(\pi \wedge \sigma)$. In this manner, the sixteen atomic binary logical operations are defined on partitions using the ditset method. But, as already noted, there are many other binary operations constructed from compound formulas of the atomic operations, e.g., $\neg (\sigma \wedge \tau)$ and $\neg \sigma \vee \neg \tau$, that could just as well be called "logical." Since there are a countable number of finite compound formulas, it is not at present known if there are only a finite number of binary logical operations on partitions. However, there are some results if we focus on the sixteen atomic logical operations. Four of the operations, the join, meet, implication, and nand, suffice to define the other atomic operations.

Lemma 5 *For any subsets $S, T \subseteq U \times U$,* int $[S \cap T] =$ int [int $(S) \cap$ int (T)].

Proof: Since int $(S) \subseteq S$ and int $(T) \subseteq T$, int [int $(S) \cap$ int (T)] $\subseteq$ int $[S \cap T]$. Conversely, $S \cap T \subseteq S, T$ so int $(S \cap T) \subseteq$ int $(S) \cap$ int (T) and since int $(S \cap T)$ is open, int $[S \cap T] \subseteq$ int [int $(S) \cap$ int (T)]. $\square$

The four atomic subset areas, $S \cap T$, $S \cap T^c$, $S^c \cap T$, and $S^c \cap T^c$, correspond to the disjunctive normal form in subset logic. But the Lemma shows that it is conjunctive normal form (CNF) that might be more useful in partition logic–since the corresponding result for disjunctive normal form does not hold. Hence the strategy is to express each of the 15 (non-universal) regions definable in $U \times U$ from S and T as in Table 1.11 and then take $S =$ dit (σ) and $T =$ dit (τ) and take the interior of the meet and use the Lemma.[6]

---

[6]The notation for some of the binary logical operations is taken from Church [17].

15 regions Conjunctive Normal Form	Binary operation on partitions
$(S \cup T) \cap (S^c \cup T) \cap (S \cup T^c) \cap (S^c \cup T^c)$	0
$(S^c \cup T) \cap (S \cup T^c) \cap (S^c \cup T^c)$	$\sigma \overline{\vee} \tau = \neg \sigma \wedge \neg \tau$
$(S \cup T) \cap (S \cup T^c) \cap (S^c \cup T^c)$	$\tau \nLeftarrow \sigma = \sigma \wedge \neg \tau$
$(S \cup T^c) \cap (S^c \cup T^c)$	$\neg \tau = \tau \Rightarrow 0$
$(S \cup T) \cap (S^c \cup T) \cap (S^c \cup T^c)$	$\sigma \nRightarrow \tau = \neg \sigma \wedge \tau$
$(S^c \cup T) \cap (S^c \cup T^c)$	$\neg \sigma = \sigma \Rightarrow 0$
$(S \cup T) \cap (S^c \cup T^c)$	$\sigma \not\equiv \tau$
$S^c \cup T^c$	$\sigma \mid \tau$
$(S \cup T) \cap (S^c \cup T) \cap (S \cup T^c)$	$\sigma \wedge \tau$
$(S^c \cup T) \cap (T^c \cup S)$	$\sigma \equiv \tau$
$(S \cup T) \cap (S \cup T^c)$	σ
$S \cup T^c$	$\tau \Rightarrow \sigma$
$(S \cup T) \cap (S^c \cup T)$	τ
$S^c \cup T$	$\sigma \Rightarrow \tau$
$S \cup T$	$\sigma \vee \tau$

Table 1.11: Interior of Column 1 gives partition operation in Column 2

Using the Lemma, we may distribute the interior of the CNF subset across it with $S = \operatorname{dit}(\sigma)$ and $T = \operatorname{dit}(\tau)$, we get the expression for the 15 atomic operations in terms of the meet, join, implication, and nand.[7] The sixteenth operation, the constant $\mathbf{1}$, can be obtained as $\sigma \Rightarrow \sigma$ or $\tau \Rightarrow \tau$.

---

[7]There are other combinations which can be taken as primitive since the *inequivalence, symmetric difference, exclusive-or,* or *xor* $\sigma \not\equiv \tau$ can be used to define the nand operation: $((\sigma \vee \tau) \Rightarrow (\sigma \not\equiv \tau)) = \sigma \mid \tau$.

Binary operation	Partition CNF for 15 binary operations
0	$= (\sigma \vee \tau) \wedge (\sigma \Rightarrow \tau) \wedge (\tau \Rightarrow \sigma) \wedge (\sigma|\tau)$
$\sigma \overline{\vee} \tau = \neg\sigma \wedge \neg\tau$	$= (\sigma \Rightarrow \tau) \wedge (\tau \Rightarrow \sigma) \wedge (\sigma|\tau)$
$\tau \not\Leftarrow \sigma = \sigma \wedge \neg\tau$	$= (\sigma \vee \tau) \wedge (\tau \Rightarrow \sigma) \wedge (\sigma|\tau)$
$\neg\tau = \tau \Rightarrow 0$	$= (\tau \Rightarrow \sigma) \wedge (\sigma|\tau)$
$\sigma \not\Leftarrow \tau = \neg\sigma \wedge \tau$	$= (\sigma \vee \tau) \wedge (\sigma \Rightarrow \tau) \wedge (\sigma|\tau)$
$\neg\sigma = \sigma \Rightarrow 0$	$= (\sigma \Rightarrow \tau) \wedge (\sigma|\tau)$
$\sigma \not\equiv \tau$	$= (\sigma \vee \tau) \wedge (\sigma|\tau)$
$\sigma|\tau$	$= \sigma|\tau$
$\sigma \wedge \tau$	$= (\sigma \vee \tau) \wedge (\sigma \Rightarrow \tau) \wedge (\tau \Rightarrow \sigma)$
$\sigma \equiv \tau$	$= (\sigma \Rightarrow \tau) \wedge (\tau \Rightarrow \sigma)$
σ	$= (\sigma \vee \tau) \wedge (\tau \Rightarrow \sigma)$
$\tau \Rightarrow \sigma$	$= \tau \Rightarrow \sigma$
τ	$= (\sigma \vee \tau) \wedge (\sigma \Rightarrow \tau)$
$\sigma \Rightarrow \tau$	$= \sigma \Rightarrow \tau$
$\sigma \vee \tau$	$= \sigma \vee \tau$

Table 1.12: Distributing interior across intersections gives partition CNF.

In addition to the two constants 0 and 1, and the four unary operations σ, τ, $\neg\sigma$, and $\neg\tau$, the ten binary operations of Table 1.12 occur in natural pairs: $\Rightarrow$ and $\not\Rightarrow$, $\Leftarrow$ and $\not\Leftarrow$, $\equiv$ and $\not\equiv$, $\vee$ and $\overline{\vee}$, and $\wedge$ and $|$. The relationship between the operations in the pairs is not negation (except for the join); it is orthogonality. If one of the operations applied to σ and τ is non-zero, the other must be zero (the indiscrete partition). There is another pairing of the operations by complemention-duality.

1.4.5   The sixteen binary logical operations on equivalence relations

For every operation on partitions or partition relations, there is a complementation-dual operation on equivalence relations. The lattice of partitions on U enriched by implication gives the partition algebra $\Pi(U)$. It is isomorphic to the algebra of open subsets (partition relations or ditsets) $\mathcal{O}(U \times U)$ of $U \times U$ under the mapping $\pi \leftrightarrow \mathrm{dit}(\pi)$. Thus $\mathcal{O}(U \times U)$ is anti-isomorphic (under order-reversing complementation) to the algebra of closed subsets or equivalence relations $\mathcal{C}(U \times U)$ on $U \times U$. The order-reversing isomorphism between $\Pi(U)$ and $\mathcal{C}(U \times U)$ is $\pi \leftrightarrow \mathrm{indit}(\pi)$. In order to keep notation analogous, we will refer to the equivalence relation $\mathrm{indit}(\pi)$ as π^d, the complementation-dual to π.

One could define the logical operations on equivalence relations directly but it is more convenient to define them by duality. The top of the dual algebra of equivalence relations (ordered by inclusion) is $\hat{1}_U = 0_U^d = \text{indit}(0_U) = U \times U$, the universal equivalence relation and the bottom is $\hat{0}_U = 1_U^d = \text{indit}(1_U) = \Delta$, the diagonal.

The operations on equivalence relations or inditsets is defined as the inditset of the dual (i.e., complementation-dual) operation on ditsets.

- Join: $\text{indit}(\pi) \vee \text{indit}(\sigma) = \text{indit}(\pi \wedge \sigma) = \overline{(\text{dit}(\pi) \cap \text{dit}(\sigma))}$ which in the d-superscript notation is: $\pi^d \vee \sigma^d = (\pi \wedge \sigma)^d = \text{indit}(\pi \wedge \sigma)$;

- Meet: $\text{indit}(\pi) \wedge \text{indit}(\sigma) = \text{indit}(\pi \vee \sigma) = \text{indit}(\pi) \cap \text{indit}(\sigma)$ or $\pi^d \wedge \sigma^d = (\pi \vee \sigma)^d = \text{indit}(\pi \vee \sigma)$; and

- Difference: $\text{indit}(\pi) - \text{indit}(\sigma) = \text{indit}(\sigma \Rightarrow \pi) = \overline{(\text{dit}(\sigma) \cap \text{indit}(\pi))}$, or: $\pi^d - \sigma^d = (\sigma \Rightarrow \pi)^d = \text{indit}(\sigma \Rightarrow \pi)$.

The dual of the implication for subsets $S \supset T = S^c \cup T$ by complementation is: $(S^c \cup T)^c = S \cap T^c = S - T$ hence the complementation-dual of implication is difference. That gives the definition of the dual algebra of equivalence relations $C(U \times U)$ with the constants of $\hat{0}_U$ and $\hat{1}_U$ and the four primitive operations of meet, join, and difference.

For any compound formula φ using the four atomic partition operations, the dual equivalence relation φ^d is proven by standard methods to be indit (φ).

Proposition 3 $\varphi^d = \text{indit}(\varphi)$.

Proof: The proof uses induction over the complexity of the formulas [where complexity is defined in the standard way in propositional logic]. If φ is one of the constants 0_U or 1_U, then the proposition holds since: $0_U^d = \hat{1}_U = \text{indit}(0_U)$ and $1_U^d = \hat{0}_U = \text{indit}(1_U)$. If $\varphi = \alpha$ is atomic, then it is true by the definition: $\sigma^d = \text{indit}(\sigma)$. If φ is a compound formula then the main connective in φ is one of the four primitive partition operations and the main connective in φ^d is one of the four primitive equivalence relation operations. Consider the case: $\varphi = \pi \wedge \sigma$ so that $\varphi^d = \pi^d \vee \sigma^d$. By the induction hypothesis, $\pi^d = \text{indit}(\pi)$ and $\sigma^d = \text{indit}(\sigma)$, and by the definition of the equivalence relation join: $\varphi^d = \pi^d \vee \sigma^d = \text{indit}(\pi) \vee \text{indit}(\sigma) = \overline{\{\text{indit}(\pi) \cup \text{indit}(\sigma)\}} = \text{indit}(\varphi)$. The other three cases proceed in a similar manner. $\square$

Substituting the dual of 0_U, namely $\hat{1}_U$, for π^d gives the dual of negation, as: $\hat{1}_U - \sigma^d$ which might be called "non-σ^d" and denoted as $-\sigma^d$.[8] The Common-Dits Theorem was equivalent to the result that if the union of two equivalence

---

[8]See Lawvere's discussion of "non-A" in co-Heyting algebras [68].

relations was the universal relation, then one of the equivalence relations was the universal relation $U \times U$. That is, dit $(\pi) \cap$ dit $(\sigma) = ($indit $(\pi) \cup$ indit $(\sigma))^c$ so if dit $(\pi) \cap$ dit $(\sigma) = \emptyset$, then the Common-dits Theorem says that dit $(\pi) = \emptyset$ or dit $(\sigma) = \emptyset$, i.e., if indit $(\pi) \cup$ indit $(\sigma) = U \times U$, then indit $(\pi) = U \times U$ or indit $(\sigma) = U \times U$. In the partition case, this implies that if $\sigma \neq \mathbf{0}_U$, then $\sigma \Rightarrow \mathbf{0}_U = \neg \sigma = \mathbf{0}_U$. In the equivalence relation case, if $\sigma^d \neq \hat{\mathbf{1}}_U$, then $\hat{\mathbf{1}}_U - \sigma^d = -\sigma^d = \hat{\mathbf{1}}_U$. Thus subtracting any equivalence relation less than $U \times U$ from $U \times U$ leaves $U \times U$. But subtracting $U \times U$ from itself leaves $\hat{\mathbf{0}}_U$.

The key result, derivable from the Common-dits Theorem is important enough to get its own proof.

Proposition 4 *The RST-closure* $\overline{\text{dit}\,(\sigma)}$ *of non-empty ditset* dit (σ) *is the universal equivalence relation* $U \times U$.

Proof: Since dit (σ) is non-empty, there exists $C, C' \in \sigma$ with $C \neq C'$ and for any $u \in C$ and $u' \in C'$, $(u, u') \in$ dit (σ). The RST-closure adds the diagonal Δ to the set so the only pairs that need to be added to make $U \times U$ are the pairs of indit (σ) such as $(u, u'') \in C$.

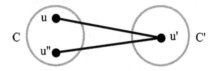

Figure 1.12: Point u connected to u'' by transitivity in RST-closure of dit (σ).

As in Figure 1.12, (u, u') and $(u', u'') \in$ dit (σ) so by transitivity, any pair $(u, u'') \in$ indit (σ) is also included in the RST-closure $\overline{\text{dit}\,(\sigma)}$ so it is $U \times U$. $\square$

In the case of $\sigma^d \neq \hat{\mathbf{1}}_U$ implying $-\sigma^d = \hat{\mathbf{1}}_U$, $\sigma^d = $ indit $(\sigma) \neq$ indit $(\mathbf{0}_U) = \hat{\mathbf{1}}_U$ means that dit $(\sigma) \neq \emptyset$ so the closure $\overline{(\text{dit}\,(\sigma) \cap \text{indit}\,(\mathbf{0}_U))} = \overline{\text{dit}\,(\sigma)}$ which by the Proposition is $U \times U =$ indit $(\mathbf{0}_U) = \hat{\mathbf{1}}_U$. When $\sigma = \mathbf{0}_U$, then dit $(\sigma) = \emptyset$ whose RST-closure is the diagonal $\Delta =$ indit $(\mathbf{1}_U) = \mathbf{1}_U^d = \hat{\mathbf{0}}_U$, the bottom of the algebra $\mathcal{C}\,(U \times U)$ of equivalence relations.

The Lemma int $[A \cap B] =$ int $[$int $(A) \cap$ int $(B)]$ for $A, B \subseteq U \times U$ could also be expressed using the closure operation as $\overline{[A \cup B]} = \overline{[\overline{A} \cup \overline{B}]}$. Hence the conjunctive normal form treatment of the 15 binary operations on partitions in terms of the operations of $\vee, \wedge, \Rightarrow$, and $|$ dualizes to the disjunctive normal form (DNF) treatment of the 15 (dual) binary operations on equivalence relations in terms of the dual operations $\wedge, \vee, -$, and $\triangledown$, which are the primitive operations in the algebra of equivalence relations $\mathcal{C}\,(U \times U)$.

The previous pair of tables giving the CNF treatment of the 15 operations on partitions complementation-dualize to give two similar tables for the DNF treatment of the 15 non-zero operations on equivalence relations. In the Table 1.13, let $S' = \text{indit}(\sigma)$ and $T' = \text{indit}(\tau)$ where $()^c$ is complementation in $U \times U$. We have also taken the liberty of writing the "converse non-implication" operation as the difference operation on both equivalence relations and partitions: $\tau^d - \sigma^d = \sigma^d \nRightarrow \tau^d$ and $\tau - \sigma = \sigma \nRightarrow \tau$.

15 regions Disjunctive Normal Form	Bin. op. on eq. rel.	Dual to
$S'^c \cap T'^c$	$\sigma^d \overline{\vee} \tau^d$	$\sigma \mid \tau$
$S' \cap T'^c$	$\sigma^d - \tau^d$	$\tau \Rightarrow \sigma$
$(S' \cap T'^c) \cup (S'^c \cap T'^c)$	$-\tau^d$	$\neg \tau$
$S'^c \cap T'$	$\tau^d - \sigma^d$	$\sigma \Rightarrow \tau$
$(S'^c \cap T') \cup (S'^c \cap T'^c)$	$-\sigma^d$	$\neg \sigma$
$(S'^c \cap T') \cup (S' \cap T'^c)$	$\sigma^d \not\equiv \tau^d$	$\sigma \equiv \tau$
$(S'^c \cap T') \cup (S'^c \cap T'^c) \cup (S' \cap T'^c)$	$\sigma^d \mid \tau^d$	$\sigma \overline{\vee} \tau$
$S' \cap T'$	$\sigma^d \wedge \tau^d$	$\sigma \vee \tau$
$(S' \cap T') \cup (S'^c \cap T'^c)$	$\sigma^d \equiv \tau^d$	$\sigma \not\equiv \tau$
$(S' \cap T') \cup (S' \cap T'^c)$	σ^d	σ
$(S' \cap T') \cup (S' \cap T'^c) \cup (S'^c \cap T'^c)$	$\tau^d \Rightarrow \sigma^d$	$\sigma - \tau$
$(S' \cap T') \cup (S'^c \cap T')$	τ^d	τ
$(S'^c \cap T') \cup (S'^c \cap T'^c) \cup (S' \cap T')$	$\sigma^d \Rightarrow \tau^d$	$\tau - \sigma$
$(S' \cap T') \cup (S' \cap T'^c) \cup (S'^c \cap T')$	$\sigma^d \vee \tau^d$	$\sigma \wedge \tau$
$(S' \cap T') \cup (S' \cap T'^c) \cup (S'^c \cap T') \cup (S'^c \cap T'^c)$	$\hat{1}$	$\mathbf{0}$

Table 1.13: RST-closure of Column 1 gives equivalence relation binary operation in Column 2

The CNF expression for the partition symmetric difference or inequivalence is: $\sigma \not\equiv \tau = (\sigma \vee \tau) \wedge (\sigma \mid \tau)$ so that:

$$\text{dit}(\sigma \not\equiv \tau) = \text{int}\left[\text{int}(\text{dit}(\sigma) \cup \text{dit}(\tau)) \cap \text{int}(\text{dit}(\sigma)^c \cup \text{dit}(\tau)^c)\right]$$
$$= \text{int}\left[(\text{dit}(\sigma) \cup \text{dit}(\tau)) \cap (\text{dit}(\sigma)^c \cup \text{dit}(\tau)^c)\right].$$

Taking complements yields:

$$\text{indit}\,(\sigma \not\equiv \tau) = \overline{[(\text{indit}\,(\sigma) \cap \text{indit}\,(\tau)) \cup (\text{indit}\,(\sigma)^c \cap \text{indit}\,(\tau)^c)]}$$

$$= \overline{\left[\overline{(\text{indit}\,(\sigma) \cap \text{indit}\,(\tau))} \cup \overline{(\text{indit}\,(\sigma)^c \cap \text{indit}\,(\tau)^c)}\right]}$$

$$= \overline{[(\sigma^d \wedge \tau^d) \cup (\sigma^d \triangledown \tau^d)]}$$

$$= \left(\sigma^d \wedge \tau^d\right) \vee \left(\sigma^d \triangledown \tau^d\right)$$

$$= \sigma^d \equiv \tau^d.$$

Thus the equivalence $\sigma^d \equiv \tau^d$ of equivalence relations has the disjunctive normal form: $\sigma^d \equiv \tau^d = (\sigma^d \wedge \tau^d) \vee (\sigma^d \triangledown \tau^d)$ in the "dual" logic of equivalence relations. The disjunctive normal forms for the 15 operations on equivalence relations is given in the following Table 1.14.

Binary operation	Equivalence relation DNF for 15 binary operations
$\sigma^d \triangledown \tau^d$	$= \sigma^d \triangledown \tau^d$
$\sigma^d - \tau^d$	$= \sigma^d - \tau^d$
$-\tau^d$	$= \left(\sigma^d - \tau^d\right) \vee \left(\sigma^d \triangledown \tau^d\right)$
$\tau^d - \sigma^d$	$= \tau^d - \sigma^d$
$-\sigma^d$	$= \left(\tau^d - \sigma^d\right) \vee \left(\sigma^d \triangledown \tau^d\right)$
$\sigma^d \not\equiv \tau^d$	$= \left(\tau^d - \sigma^d\right) \vee \left(\sigma^d - \tau^d\right)$
$\sigma^d | \tau^d$	$= \left(\tau^d - \sigma^d\right) \vee \left(\sigma^d \triangledown \tau^d\right) \vee \left(\sigma^d - \tau^d\right)$
$\sigma^d \wedge \tau^d$	$= \sigma^d \wedge \tau^d$
$\sigma^d \equiv \tau^d$	$= \left(\sigma^d \wedge \tau^d\right) \vee \left(\sigma^d \triangledown \tau^d\right)$
σ^d	$= \left(\sigma^d \wedge \tau^d\right) \vee \left(\sigma^d - \tau^d\right)$
$\tau^d \Rightarrow \sigma^d$	$= \left(\sigma^d \wedge \tau^d\right) \vee \left(\sigma^d - \tau^d\right) \vee \left(\sigma^d \triangledown \tau^d\right)$
τ^d	$= \left(\sigma^d \wedge \tau^d\right) \vee \left(\tau^d - \sigma^d\right)$
$\sigma^d \Rightarrow \tau^d$	$= \left(\sigma^d \wedge \tau^d\right) \vee \left(\tau^d - \sigma^d\right) \vee \left(\sigma^d \triangledown \tau^d\right)$
$\sigma^d \vee \tau^d$	$= \left(\sigma^d \wedge \tau^d\right) \vee \left(\sigma^d - \tau^d\right) \vee \left(\tau^d - \sigma^d\right)$
$\widehat{1}$	$= \left(\sigma^d \wedge \tau^d\right) \vee \left(\sigma^d - \tau^d\right) \vee \left(\tau^d - \sigma^d\right) \vee \left(\sigma^d \triangledown \tau^d\right)$

Table 1.14: Distributing closure across unions gives equivalence relation DNFs.

We should keep different notions of duality separate. The logic of partitions is dual to the logic of subsets in the sense that partitions are category-theoretically dual to subsets. But the logic of equivalence relations is dual to the logic of partitions in the sense of complementation duality–since equivalence relations are the complements to partition relations in $U \times U$. The duality between the Heyting algebra of open subsets is the complementation-dual to the co-Heyting algebra of closed subsets. That duality is analogous to the duality between the algebra of partitions $\Pi(U)$ (or partition relations $\mathcal{O}(U \times U)$) and the algebra of equivalence relations $\mathcal{C}(U \times U)$–except that the RST-closure operation is not topological (union of two RST-closed subsets is not necessarily RST-closed).

Chapter 2

Partition tautologies

2.1 Subset, truth-table, and partition tautologies

We consider formulas composed using the operations of $\vee$, $\wedge$, and $\Rightarrow$ as well as the top and bottom constants so that all formulas can be interpreted as being in subset logic, propositional logic, or partition logic (as well as intuitionistic propositional logic or Heyting algebras). A *subset tautology* is a formula so that no matter what subsets of any universe U ($|U| \geq 1$) are substituted for the atomic variables, the formulas evaluates (using the subset interpretation of the operations) to the universe set U. A *truth-table tautology* is a formula so that no what truth values T or F are substituted for the atomic variables, the whole formula evaluates (using the truth-table definitions of the operations) to T.

All subset tautologies are truth-table tautologies since one can take $U = 1$, the one element set so that $\wp\left(1\right) = \{\emptyset, 1\}$ with $\emptyset$ taken as F and 1 as T. To see the converse, consider a subset formula with m different atomic variables for which can be substituted m subsets of U. Those m subsets define 2^m atomic areas in a Venn diagram which correspond to the 2^m rows in the truth table for the formula. Each element $u \in U$ must belong to one of the atomic areas so interpret $u \in S$ as T assigned to a subset S and $u \notin S$ as F assigned to a subset S. Then if the formula is a truth-table tautology, e.g. Table 2.1, all the T's in the final column will indicate that any $u \in U$ is included in the subset constructed from the formula so that subset must be U, i.e., a truth-table tautology is a subset tautology.

σ	ρ	$\sigma \Rightarrow \rho$	$\sigma \wedge (\sigma \Rightarrow \rho)$	$[\sigma \wedge (\sigma \Rightarrow \rho)] \Rightarrow \rho$
T	T	T	T	T
T	F	F	F	T
F	T	T	F	T
F	F	T	F	T

Table 2.1: Truth-table tautology of modus ponens

Then reinterpreting the T's and F's as indicating that a $u \in U$ belongs to an atomic area or not, gives Table 2.2 showing that a truth-table tautology is also a subset tautology, i.e., no matter what atomic area an element $u \in U$ belongs to, it belongs to the subset generated by the final formula for a truth-table tautology so that subset is U.

$u \in S$	$u \in R$	$u \in S \supset R$	$u \in S \cap (S \supset R)$	$u \in [S \cap (S \supset R)] \supset R$
T	T	T	T	T
T	F	F	F	T
F	T	T	F	T
F	F	T	F	T

Table 2.2: Truth-table tautology reinterpreted as showing the formula is a subset tautology.

As mentioned previously, one of the reasons for the retarded development of the logic of partitions, even though the duality between partitions and subsets has been well-known since the middle of the twentieth century, is that subset logic is almost always presented as just propositional logic where validities or tautologies are defined as truth-table tautologies and subset tautologies are ignored.

A 'partition tautology' is defined just like a subset tautology with partitions on U substituted for subsets of U. That is, a *partition tautology* is a formula such that no matter what partitions on a universe U ($|U| \geq 2$), the formula evaluates using the partition operations to the top, the discrete partition $\mathbf{1}_U$. It is also useful to define a *weak partition tautology* as a formula that under the same substitutions never evaluates to the bottom, the indiscrete partition $\mathbf{0}_U$. Then a partition tautology is obviously a weak one too. Moreover, it is easily seen that:

φ is a weak partition tautology iff $\neg\neg\varphi$ is a partition tautology.

The law of excluded middle formula $\sigma \vee \neg\sigma$ is the simplest example of a subset tautology that is also a weak partition tautology but not a ('strong') partition tautology.

The simplest non-degenerate partition algebra is $\Pi(2)$, where $2 = \{0,1\}$ ("the" two-element set), the algebra with only two partitions, $\mathbf{0}_2 = \{\{0,1\}\}$ and $\mathbf{1}_2 = \{\{0\},\{1\}\}$. The simplest non-degenerate Boolean subset algebra is $\wp(1)$ where 1 is "the" one-element set which has only two subsets, $\emptyset$ and 1. The results of any set of partition operations on $\Pi(2)$ can be described by a 'truth table' since there are only two partitions. For instance, the partition operations on the modus ponens formula are given in Table 2.3.

σ	ρ	$\sigma \Rightarrow \rho$	$\sigma \wedge (\sigma \Rightarrow \rho)$	$[\sigma \wedge (\sigma \Rightarrow \rho)] \Rightarrow \rho$
$\mathbf{1}_2$	$\mathbf{1}_2$	$\mathbf{1}_2$	$\mathbf{1}_2$	$\mathbf{1}_2$
$\mathbf{1}_2$	$\mathbf{0}_2$	$\mathbf{0}_2$	$\mathbf{0}_2$	$\mathbf{1}_2$
$\mathbf{0}_2$	$\mathbf{1}_2$	$\mathbf{1}_2$	$\mathbf{0}_2$	$\mathbf{1}_2$
$\mathbf{0}_2$	$\mathbf{0}_2$	$\mathbf{1}_2$	$\mathbf{0}_2$	$\mathbf{1}_2$

Table 2.3: 'Truth table' for modus ponens formula in $\Pi(2)$.

The basic point is that the partition operations on $\Pi(2)$ have the same 'truth tables' as the propositional or subset operations in $\wp(1)$ so the two algebras are isomorphic.[1]

$$\wp(1) \cong \Pi(2)$$

In particular, any formula that is a weak partition tautology would never evaluate to $\mathbf{0}_2$ so it must always evaluate to $\mathbf{1}_U$ in $\Pi(2)$ which is isomorphic, in turn, to $\wp(1)$ so it must always evaluate to 1 or T in the truth table for the formula. Hence we have proven:

Proposition 5 *Any weak partition tautology is a truth-table tautology.* □

Corollary 2 *Any partition tautology is a truth-table tautology.* □

For an example of truth-table tautology that is not a weak partition tautology, consider the simplest example of non-distributivity where $\tau = \{\{a,b\},\{c\}\}$, $\sigma = \{\{a,c\},\{b\}\}$, and $\pi = \{\{a\},\{b,c\}\}$ and the formula is:

$$[\tau \wedge (\pi \vee \sigma)] \Rightarrow [(\tau \wedge \pi) \vee (\tau \wedge \sigma)].$$

---

[1]Since $\Pi(2)$ has essentially only one distinction (between 0 and 1), when Spencer-Brown [102] defined the algebra giving "the distinction," it turned out to be the Boolean algebra $\wp(1)$ [6].

That formula is an ordinary tautology but for those partitions substituted for the variables, the result is $\mathbf{0}_U$ so it is not a weak partition tautology. Thus the weak partition tautologies (and thus the partition tautologies) are properly contained in the truth-table tautologies.

When comparing partition tautologies (weak or not), there is no inclusion either way with the valid formulas of intuitionistic propositional logic (or Heyting algebras). The accumulation formula, $\sigma \Rightarrow (\pi \Rightarrow (\pi \wedge \sigma))$, is valid in both classical and intuitionistic logic but not in partition logic. The law of excluded middle, $\sigma \vee \neg\sigma$, is a weak partition tautology, and the weak law of excluded middle, $\neg\sigma \vee \neg\neg\sigma$, is a partition tautology that is not intuitionistically valid.

2.2 The finite model property

If a formula φ is not valid according to subset or partition semantics, then a universe and subsets or partitions that do not evaluation to the top would be a *countermodel*. If a formula is not a subset (or truth-table) tautology, then it suffices to take $U = 1$ to give a subset countermodel, and in view of the isomorphism, $\wp(1) \cong \Pi(2)$, it suffices to take $U = 2$ to have a partition countermodel. But what about formulas that are subset tautologies but not partition tautologies? For subsets, it suffices to take $|U| = 1$ to find countermodels. Is there an n such that for all $|U| \leq n$, if φ evaluates to $\mathbf{1}_U$ for all partitions on U, then φ is a partition tautology for any U? The following result, adapted from lattice theory, shows that there is no such n.

Proposition 6 *There is no n such that if any φ has no partition countermodel on any universe U with $|U| \leq n$, then φ is a partition tautology.*

Proof: Consider any fixed $n \geq 2$. We use the standard device of a "universal disjunction of equations" [49, p. 316] to construct a formula ω_n that evaluates to $\mathbf{1}_U$ for any substitutions of partitions on U with $|U| \leq n$ and yet the formula is not a partition tautology. Let B_n be the Bell number, the number of partitions on a set U with $|U| = n$. Take the atomic variables to be π_i for $i = 0, 1, ..., B_n$ so that there are $B_n + 1$ atomic variables. Let ω_n be the join of all the equivalences between distinct atomic variables:

$$\omega_n = \bigvee \{\pi_i \equiv \pi_j : 0 \leq i < j \leq B_n\}.$$

Then for any substitution of partitions on U where $|U| \leq n$ for the atomic variables, there is, by the pigeonhole principle, some "disjunct" $\pi_i \equiv \pi_j = (\pi_i \Rightarrow \pi_j) \wedge (\pi_j \Rightarrow \pi_i)$ which has the same partition substituted for the two variables so the disjunct evaluates to $\mathbf{1}_U$ and thus the join ω_n evaluates to $\mathbf{1}_U$. Thus ω_n evaluates to $\mathbf{1}_U$ for any substitutions of partitions on any U where $|U| \le n$. To see that ω_n is not a partition tautology, take $U = \{0, 1, ..., B_n\}$ and let π_i be the atomic partition which has i as a singleton and all the other elements of U as a block, i.e., $\pi_i = \{\{i\}, \{0, 1, ..., i-1, i+1, ..., B_n\}\}$. Then $\pi_i \Rightarrow \pi_j = \pi_j$ and $\pi_j \wedge \pi_i = \mathbf{0}_U$ so that $\omega_n = \mathbf{0}_U$ for that substitution and thus ω_n is not even a weak partition tautology. □

This result leaves open the question of whether or not partition logic has the finite model property. That is, if φ is not a partition tautology, then does it have a countermodel with a finite U?

If π is a partition on U and $S \subseteq U$, then $\pi \upharpoonright S = \{B \cap S \neq \emptyset : B \in \pi\}$ is a partition on S. Since join, meet, and implication are a complete set of connectives for ordinary propositional logic, we will consider formulas in that language.

Lemma 6 *Let φ be a formula (always finite) involving the basic connectives, join, meet, and implication, and the two constants $\mathbf{0}_U$ and $\mathbf{1}_U$. Then for any partitions $\pi_1, ..., \pi_n$ on a universe set U substituted for the atomic variables, if the partition, $\varphi(\pi_1, ..., \pi_n)$ has a proper indit (u, u') (i.e., $u \neq u'$), then the indit only involves a finite set of elements $U' \subseteq U$ such that (u, u') is also an indit of $\varphi(\pi_1 \upharpoonright U', ..., \pi_n \upharpoonright U')$.*

Proof: The proof uses induction over the complexity of the formula φ. At the atomic level, if (u, u') is an indit of a partition (including $\mathbf{0}_U$), then it is also an indit of the partition cut down to any finite U' containing u and u'. Assuming the hypothesis for formulas of lesser complexity, suppose that $\varphi = \sigma * \pi$ where $*$ is one of the three basic connectives.

- Join: indit $(\sigma \vee \pi) = $ indit $(\sigma) \cap$ indit (π) so an indit $(u, u') \in$ indit $(\sigma \vee \pi)$ only uses one indit in σ and π;

- Meet: indit $(\sigma \wedge \pi) = \overline{(\text{indit}(\sigma) \cup \text{indit}(\pi))}$ so an indit in the closure requires a finite chain of links drawn from indit (σ) and indit (π) and each of those indits uses only a finite set of elements from U by the induction hypothesis; and

- Implication: dit $(\sigma \Rightarrow \pi) = $ int $[\text{dit}(\sigma)^c \cup \text{dit}(\pi)]$ so indit $(\sigma \Rightarrow \pi) = \overline{(\text{dit}(\sigma) \cap \text{indit}(\pi))}$ and thus it only involves a finite chain of indits for π which, in turn, uses only a finite set of elements from U.

Since a finite union of finite sets is finite, there is a finite $U' \subseteq U$ so that $(u, u') \in$ indit $(\varphi(\pi_1 \upharpoonright U', ..., \pi_n \upharpoonright U'))$. □

Proposition 7 (Finite model property) *If a formula φ is not a partition tautology, then it always has a finite countermodel.*

Proof: If φ is not a partition tautology, then there is a universe U and partitions $\pi_1, ..., \pi_n$ on U such that when substituted for the atomic variables of φ, then the partition $\varphi(\pi_1, ..., \pi_n) \neq 1_U$. Hence that partition contains a non-singleton block $\{u, u', ...\}$. By the lemma, the indit (u, u') only involves a finite set $U' \subseteq U$ of elements so that (u, u') is also an indit of $\varphi(\pi_1 \upharpoonright U', ..., \pi_n \upharpoonright U')$ and thus the formula φ has a finite countermodel. $\qquad\square$

2.3   Generating partition tautologies using the Boolean core $\mathcal{B}[\pi, 1_U]$

For any fixed partition π on U, a π-regular partition in $\Pi(U)$ is a partition of the form $\sigma \Rightarrow \pi$ for any σ. We previously showed that the π-regular partitions form a Boolean algebra (BA) $\mathcal{B}[\pi, 1_U]$ where the partition operations of join, meet, and implication also act as the corresponding BA operations in $\mathcal{B}[\pi, 1_U]$ and where π is the bottom (or zero element) and 1_U is the top of the BA. Hence any classical (= subset = truth-table) tautology, e.g., $\sigma \vee (\sigma \Rightarrow 0)$, expressed in the language of join, meet, implication, and the two constants, when formulated in $\mathcal{B}[\pi, 1_U]$ would always evaluate to 1_U using the BA operations when any π-regular elements are substituted for the variables. But the BA operations are the same as the corresponding partition operations so classical tautologies in $\mathcal{B}[\pi, 1_U]$ yield partition tautologies.

  This connection can be formalized by converting any classical tautology into the corresponding tautology in $\mathcal{B}[\pi, 1_U]$ and thus into a partition tautology. Given any classical formula using the connectives of $\vee, \wedge, \Rightarrow$ and the constants of 0 and 1, its *single π-negation transform* is obtained by replacing each atomic variable σ by its single π-negation $\overset{\pi}{\neg}\sigma = \sigma \Rightarrow \pi$ and by replacing the constant 0 by π and 1 by 1_U. The binary operations $\vee, \wedge$, and $\Rightarrow$ all remain the same. For instance, the single π-negation transform of the excluded middle formula $\sigma \vee \neg\sigma = \sigma \vee (\sigma \Rightarrow 0)$ is the weak excluded middle formula for π-negation:

$$(\sigma \Rightarrow \pi) \vee ((\sigma \Rightarrow \pi) \Rightarrow \pi) = \overset{\pi}{\neg}\sigma \vee \overset{\pi}{\neg}\overset{\pi}{\neg}\sigma.$$

Then the single π-negation transformed formula is a tautology in the BA $\mathcal{B}[\pi, 1_U]$, so it is also a partition tautology. This formula is also an example of a partition tautology that is not a valid formula of intuitionistic logic (either for $\pi = 0_U$ or in general).

Proposition 8 *The single π-negation transform of any classical tautology is a partition tautology.* □

This process can be repeated using double negation. The *double π-negation transform* of a classical formula using the connectives of $\vee$, $\wedge$, $\Rightarrow$ and the constants of 0 and 1 is obtained by replacing any atomic variable σ by its double π-negation $\overset{\pi}{\neg}\overset{\pi}{\neg}\sigma = (\sigma \Rightarrow \pi) \Rightarrow \pi$ and by replacing the constant 0 by π and 1 by $\mathbf{1}_U$. The binary operations $\vee$, $\wedge$, and $\Rightarrow$ all remain the same.

Proposition 9 *The double π-negation transform of any classical tautology is a partition tautology.* □

The double π-negation transform of excluded middle is the formula $\overset{\pi}{\neg}\overset{\pi}{\neg}\sigma \vee \overset{\pi}{\neg}\overset{\pi}{\neg}\overset{\pi}{\neg}\sigma$. Since the π-negation has the effect of flipping the π-blocks B back and forth being locally equal to $\mathbf{0}_B$ or $\mathbf{1}_B$ (i.e., from being whole to being discretized), it is clear that $\overset{\pi}{\neg}\sigma = \overset{\pi}{\neg}\overset{\pi}{\neg}\overset{\pi}{\neg}\sigma$ so the formula $\overset{\pi}{\neg}\overset{\pi}{\neg}\sigma \vee \overset{\pi}{\neg}\overset{\pi}{\neg}\overset{\pi}{\neg}\sigma$ is equivalent to $\overset{\pi}{\neg}\overset{\pi}{\neg}\sigma \vee \overset{\pi}{\neg}\sigma$.

The BA of regular elements is not the only idea that can be usefully transplanted from intuitionistic logic to partition logic. The partition analogue of the Gödel transform [46] that produces an intuitionistic validity from each classical tautology can be constructed in the following manner. For any classical formula φ in the language of $\vee$, $\wedge$, and $\Rightarrow$ as well as 0 and 1, we define the *Gödel π-transform* φ_π^g of the formula as follows:

- If φ is atomic, then $\varphi_\pi^g = \varphi \vee \pi$;

- If $\varphi = 0$, then $\varphi_\pi^g = \pi$, and if $\varphi = 1$, then $\varphi_\pi^g = \mathbf{1}_U$;

- If $\varphi = \sigma \vee \tau$, then $\varphi_\pi^g = \sigma_\pi^g \vee \tau_\pi^g$;

- If $\varphi = \sigma \Rightarrow \tau$, then $\varphi_\pi^g = \sigma_\pi^g \Rightarrow \tau_\pi^g$; and

- if $\varphi = \sigma \wedge \tau$, then $\varphi_\pi^g = \overset{\pi}{\neg}\overset{\pi}{\neg}\sigma_\pi^g \wedge \overset{\pi}{\neg}\overset{\pi}{\neg}\tau_\pi^g$.

When $\pi = 0$, then we write $\varphi_0^g = \varphi^g$. We first consider the case for $\pi = 0$.

Lemma 7 *φ is a classical tautology iff φ^g is a weak partition tautology iff $\neg\neg\varphi^g$ is a partition tautology.*

Proof: The idea of the proof is that the partition operations on the Gödel 0-transform φ^g mimic the Boolean 0, 1-operations on φ if we associate the partition interpretation $\sigma^g = 0$ with the Boolean $\sigma = 0$ and $\sigma^g \neq 0$ with the

Boolean $\sigma = 1$. We proceed by induction over the complexity of the formula φ where the induction hypothesis is that: $\varphi = 1$ in the Boolean case iff $\varphi^g \neq 0$ in the partition case, which could also be stated as: $\varphi = 0$ in the Boolean case iff $\varphi^g = 0$ in the partition case. If φ is atomic, then $\varphi^g = \varphi \vee 0 = \varphi$. The Boolean assignment $\varphi = 0$ (the Boolean truth value 0) is associated with the partition assignment of $\varphi = \mathbf{0}_U$ (the indiscrete partition) and for atomic φ, $\varphi = \varphi \vee \mathbf{0}_U = \varphi^g$ so the hypothesis holds in the base case.

For the join in the Boolean case, $\varphi = \sigma \vee \tau = 1$ iff $\sigma = 1$ or $\tau = 1$. In the partition case, $\varphi^g = \sigma^g \vee \tau^g \neq \mathbf{0}_U$ iff $\sigma^g \neq \mathbf{0}_U$ or $\tau^g \neq \mathbf{0}_U$, so by the induction hypothesis, $\varphi = \sigma \vee \tau = 1$ iff $\sigma = 1$ or $\tau = 1$ iff $\sigma^g \neq \mathbf{0}_U$ or $\tau^g \neq \mathbf{0}_U$ iff $\varphi^g = \sigma^g \vee \tau^g \neq \mathbf{0}_U$.

For the implication in the Boolean case, $\varphi = \sigma \Rightarrow \tau = 0$ iff $\sigma = 1$ and $\tau = 0$. In the partition case, $\varphi^g = \sigma^g \Rightarrow \tau^g = \mathbf{0}_U$ iff $\sigma^g \neq \mathbf{0}_U$ and $\tau^g = \mathbf{0}_U$. Hence using the induction hypothesis, $\varphi = \sigma \Rightarrow \tau = 1$ iff $\sigma = 0$ or $\tau = 1$ iff $\sigma^g = \mathbf{0}_U$ or $\tau^g \neq \mathbf{0}_U$ iff $\varphi^g = \sigma^g \Rightarrow \tau^g \neq \mathbf{0}_U$.

For the meet in the Boolean case, $\varphi = \sigma \wedge \tau = 1$ iff $\sigma = 1 = \tau$. In the partition case, $\varphi^g = \neg\neg\sigma^g \wedge \neg\neg\tau^g = \mathbf{1}_U$ iff $\neg\neg\sigma^g = \mathbf{1}_U = \neg\neg\tau^g$ iff $\sigma^g \neq \mathbf{0}_U \neq \tau^g$. By the induction hypothesis, $\varphi = \sigma \wedge \tau = 1$ iff $\sigma = 1 = \tau$ iff $\sigma^g \neq \mathbf{0}_U \neq \tau^g$ iff $\varphi^g = \neg\neg\sigma^g \wedge \neg\neg\tau^g = \mathbf{1}_U$ iff $\varphi^g = \neg\neg\sigma^g \wedge \neg\neg\tau^g \neq \mathbf{0}_U$.

Thus φ is a classical tautology iff under any Boolean interpretation, $\varphi = 1$ iff for any partition interpretation, $\varphi^g \neq \mathbf{0}_U$ iff φ^g is a weak partition tautology iff $\neg\neg\varphi^g$ is a partition tautology. $\square$

In this case of $\pi = \mathbf{0}_U$, the negation $\neg\sigma = \sigma \Rightarrow \mathbf{0}_U$ is unchanged and, for atomic variables φ, $\varphi \vee \mathbf{0}_U = \varphi$ so atomic variables are left unchanged in the Gödel $\mathbf{0}_U$-transform. Hence any classical formula φ expressed in the language of $\neg$, $\vee$, and $\Rightarrow$ (excluding the meet $\wedge$) would be unchanged by the Gödel 0-transform.

Corollary 3 *For any formula φ in the language of $\neg$, $\vee$, and $\Rightarrow$ along with 0 and 1, φ is a classical tautology iff φ is a weak partition tautology iff $\neg\neg\varphi$ is a partition tautology.*

For instance, the Gödel $\mathbf{0}_U$-transform of excluded middle $\sigma \vee \neg\sigma$ is the same formula, $\sigma \vee \neg\sigma$, which is a weak partition tautology, and $\neg\neg(\sigma \vee \neg\sigma)$ is a partition tautology.

The lemma generalizes to any π in the following form.

Proposition 10 *φ is a classical tautology iff $\overset{\pi}{\neg}\overset{\pi}{\neg}\varphi_\pi^g$ is a partition tautology.*

Proof: For any fixed partition π on a universe set U, the interpretation of the Gödel π-transform φ_π^g is in the upper interval $[\pi, \mathbf{1}_U] \subseteq \Pi(U)$. The key to the generalization is the standard result [2] that the upper interval $[\pi, \mathbf{1}_U]$ can be represented as the product of the sets $\Pi(B)$ where B is a non-singleton block of π:

$$[\pi, \mathbf{1}_U] \cong \prod \{\Pi(B) : B \in \pi, \ B \text{ non-singleton}\}.$$

Once we establish that the Gödel π-transform φ_π^g can be obtained, using the isomorphism, by computing the Gödel 0-transform φ^g "component-wise" in $\Pi(B)$, then we can apply the lemma component-wise to obtain the result.

Given a partition π on U, any interpretation of an atomic φ as a partition on U can be cut down to each non-singleton block $B \in \pi$ to yield a partition on B. Then $\varphi_\pi^g = \varphi \vee \pi$ has a block $B \in \pi$ iff $\varphi_0^g = \varphi^g$ is equal to the zero $\mathbf{0}_B$ of $\Pi(B)$. Proceeding by induction over the complexity of φ, if $\varphi = \sigma \vee \tau$, then a block of $\varphi_\pi^g = \sigma_\pi^g \vee \tau_\pi^g$ is B iff B is a block of both σ_π^g and τ_π^g iff $\sigma^g = \mathbf{0}_B = \tau^g$ in $\Pi(B)$ iff $\varphi^g = \sigma^g \vee \tau^g = \mathbf{0}_B$ in $\Pi(B)$. If $\varphi = \sigma \Rightarrow \tau$, then $\varphi_\pi^g = \sigma_\pi^g \Rightarrow \tau_\pi^g$ has a block $B \in \pi$ iff σ_π^g does not have the block B and τ_π^g has the block B iff σ^g is not equal to $\mathbf{0}_B$ and τ^g is equal to $\mathbf{0}_B$ in $\Pi(B)$ iff $\varphi^g = \sigma^g \Rightarrow \tau^g = \mathbf{0}_B$ in $\Pi(B)$. If $\varphi = \sigma \wedge \tau$, then $\varphi_\pi^g = \overset{\pi}{\neg}\overset{\pi}{\neg}\sigma_\pi^g \wedge \overset{\pi}{\neg}\overset{\pi}{\neg}\tau_\pi^g$ has a block $B \in \pi$ iff both σ_π^g and τ_π^g have a block B iff $\sigma^g = \mathbf{0}_B = \tau^g$ in $\Pi(B)$ iff $\varphi^g = \neg\neg\sigma^g \wedge \neg\neg\tau^g = \mathbf{0}_B$ in $\Pi(B)$.

Hence applying the lemma component-wise, φ is a classical tautology iff φ^g never evaluates to $\mathbf{0}_B$ in $\Pi(B)$ iff B is never a block of φ_π^g iff every block $B \in \pi$ is discretized in $\overset{\pi}{\neg}\overset{\pi}{\neg}\varphi_\pi^g$, i.e., $\overset{\pi}{\neg}\overset{\pi}{\neg}\varphi_\pi^g$ is a partition tautology. $\qquad\square$

Thus the Gödel π-transform of excluded middle $\varphi = \sigma \vee (\sigma \Rightarrow 0)$ is $\varphi_\pi^g = (\sigma \vee \pi) \vee (\sigma \Rightarrow \pi)$ and $\overset{\pi}{\neg}\overset{\pi}{\neg}[(\sigma \vee \pi) \vee (\sigma \Rightarrow \pi)]$ is a partition tautology. Note that the single π-negation transform, the double π-negation transform, and the Gödel π-transform all gave different formulas starting with the classical excluded middle tautology.

2.4 Some partition tautologies

One source of some interesting results on partitions was Oystein Ore's early work on equivalence relations [79]. He defined two partitions to be *associable* if every block in their meet was a block of one or the other (or both).

Lemma 8 *The following statements are equivalent:*

1. σ and τ are associable;

---

[2]Since the partition lattice is conventionally written upside down, the usual result is stated in terms of the interval below π [49, p. 192].

2. no block of one partition just partially overlaps with a block of the other, i.e., for any $C \in \sigma$ and $D \in \tau$, if $C \cap D \neq \emptyset$, then $C \subseteq D$ or $D \subseteq C$;

3. indit $(\sigma) \cup$ indit $(\tau) = $ indit $(\sigma \wedge \tau)$, i.e., dit $(\sigma) \cap$ dit $(\tau) = $ dit $(\sigma \wedge \tau)$.

Proof: If $C \cap D \neq \emptyset$, $C - D \neq \emptyset$, and $D - C \neq \emptyset$, then a block in the meet will strictly contain C and D so it is not a block in σ or in τ so not #2 implies not #1, i.e., #1 implies #2 And if #2 holds, and $C \cap D \neq \emptyset$, then say $C \subseteq D$. Then for any other $C' \cap D \neq \emptyset$, we cannot have $D \subseteq C'$ so $C' \subseteq D$. Thus D is the exact union of the $C \in \sigma$ that overlap it so it is a block in the meet. And if $C \cap D \neq \emptyset$ and $D \subseteq C$, then by symmetry C is the exact union of $D \in \tau$ that overlap it so it is a block in the meet. Hence #1 and #2 are equivalent. Since indit $(\sigma \wedge \tau) = \overline{\text{indit} (\sigma) \cup \text{indit} (\tau)}$, if #3 is not true, then there exists $(u, u') \in$ indit (σ) and $(u', u'') \in$ indit (τ) so that (u, u'') is in the closure but not in the union. But that implies that there is a $C \in \sigma$ and $D \in \tau$ with $u \in C - D$, $u' \in C \cap D$, and $u'' \in D - C$ which negates #2 so #2 implies #3. Suppose #3 holds. For any $C \cap D \neq \emptyset$, there is a block $M \in \sigma \wedge \tau$ with $C, D \subseteq M$. But if $C \nsubseteq D$ and $D \nsubseteq C$, then there is $u \in C - D$, $u' \in C \cap D$, and $u'' \in D - C$. Then $(u, u') \in$ indit (σ) and $(u', u'') \in$ indit (τ) and the union of the inditsets is an inditset, so (u, u'') must be included in the union but it is included in neither of the inditsets in the union. Hence $C \subseteq D$ or $D \subseteq C$ so #3 implies #2. $\qquad \square$

Although Ore didn't define the π-regular partitions (not having the partition implication), the partitions in the Boolean core $\mathcal{B} [\pi, 1_U]$ are all associable since their block-wise operation follows from their 'truth table' (see Table 4.2). Ore showed that any partition joined with the meet of two associable partitions will distribute across the meet. Hence we have the following corollary for any partitions φ, σ, τ, and π.

Lemma 9 (Ore's associability theorem)

$$\varphi \vee \left(\overset{\pi}{\neg} \sigma \wedge \overset{\pi}{\neg} \tau \right) = \left(\varphi \vee \overset{\pi}{\neg} \sigma \right) \wedge \left(\varphi \vee \overset{\pi}{\neg} \tau \right).$$

Moreover, if we assume that φ is in the upper segment $[\pi, 1_U]$, then the other distributivity law holds. In proving results about π-regular partitions, it is often useful to think in terms of B-slots as pictured in Figure 2.1.

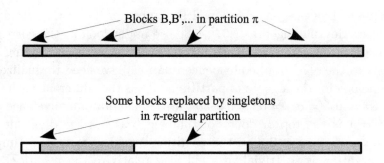

Figure 2.1: The B-slots in a π-regular partition.

Lemma 10 ("Dual" to Ore's theorem) *If* $\pi \precsim \varphi$, *then* $\varphi \wedge \left(\overset{\pi}{\neg}\sigma \vee \overset{\pi}{\neg}\tau\right) = \left(\varphi \wedge \overset{\pi}{\neg}\sigma\right) \vee \left(\varphi \wedge \overset{\pi}{\neg}\tau\right)$.

Proof: If $\pi \precsim \varphi$, then every block $J \in \varphi$ is contained in a block B of π. If the result on the left-hand side for $\left(\overset{\pi}{\neg}\sigma \vee \overset{\pi}{\neg}\tau\right)$ for that B-slot is $\mathbf{1}_B$ (i.e., the discretized version of B), then J stays a block in $\varphi \wedge \left(\overset{\pi}{\neg}\sigma \vee \overset{\pi}{\neg}\tau\right)$. If the result of $\left(\overset{\pi}{\neg}\sigma \vee \overset{\pi}{\neg}\tau\right)$ for that B is $\mathbf{0}_B$, then B is a block in the meet $\varphi \wedge \left(\overset{\pi}{\neg}\sigma \vee \overset{\pi}{\neg}\tau\right)$. On the right-hand side, if the result of $\left(\overset{\pi}{\neg}\sigma \vee \overset{\pi}{\neg}\tau\right)$ concerning that B-slot was $\mathbf{1}_B$, then either $\overset{\pi}{\neg}\sigma$ or $\overset{\pi}{\neg}\tau$ (or both) was $\mathbf{1}_B$ then either $\left(\varphi \wedge \overset{\pi}{\neg}\sigma\right)$ or $\left(\varphi \wedge \overset{\pi}{\neg}\tau\right)$ is J in part of the B-slot (and the other is J or B) so the join is J. If the B-slot in $\left(\overset{\pi}{\neg}\sigma \vee \overset{\pi}{\neg}\tau\right)$ was $\mathbf{0}_B$, then both $\left(\varphi \wedge \overset{\pi}{\neg}\sigma\right)$ and $\left(\varphi \wedge \overset{\pi}{\neg}\tau\right)$ are B and thus so is the meet. $\square$

Corollary 4 *Any* $\varphi \in [\pi, \mathbf{1}_U]$ *and any* π-*regular partition are associable.*

Proof: The meet $\varphi \wedge \overset{\pi}{\neg}\sigma$ has in the B-slot either J (along with other $J' \in \varphi$) or B so its blocks are always a block of one of the partitions in the meet. $\square$

Proposition 11 (Distributivity over the Boolean core) *If* $\pi \preceq \varphi$,

$$\varphi \vee \left(\overset{\pi}{\neg}\sigma \wedge \overset{\pi}{\neg}\tau\right) = \left(\varphi \vee \overset{\pi}{\neg}\sigma\right) \wedge \left(\varphi \vee \overset{\pi}{\neg}\tau\right)$$
$$\varphi \wedge \left(\overset{\pi}{\neg}\sigma \vee \overset{\pi}{\neg}\tau\right) = \left(\varphi \wedge \overset{\pi}{\neg}\sigma\right) \vee \left(\varphi \wedge \overset{\pi}{\neg}\tau\right). \quad \square$$

F. William Lawvere has developed a number of interesting results in the context of co-Heyting algebras ([67]; [68]) we might look at the corresponding results in partition algebras. Since he was working with co-Heyting algebras (analogous to the algebra of equivalence relations), we need to dualize some of the terminology for the algebra of partitions. Thus the "difference from 1" (e.g., $(\neg\sigma)^d = (\sigma \Rightarrow 0)^d = \sigma^d \neq 0^d = 0^d - \sigma^d$ in the algebra of equivalence relations where $0^d = \hat{1}$ is the top or "one" of that algebra) has the dual "implication to $\mathbf{0}_U$," i.e., $\neg\sigma = \sigma \Rightarrow \mathbf{0}_U$, in the partition algebra. Since the "implication to zero" negation is so trivial, we will also relativize the negation using an arbitrary π in place of $\mathbf{0}_U$.

Lawvere defined the "boundary" of an element as the meet with its negation, so dualizing we can define the π-*coboundary* of a partition σ as the join $\partial^\pi\sigma = \sigma \vee \overset{\pi}{\neg}\sigma$ with its π-negation. Thinking in terms of the B-slot in $\partial^\pi\sigma = \sigma \vee \overset{\pi}{\neg}\sigma$, if there is a $C \in \sigma$ with $B \subseteq C$, then the B-slot is $\mathbf{1}_B$ and otherwise the B-slot is the restriction $\sigma \restriction B = \{C \cap B \neq \emptyset : C \in \sigma\}$. Lawvere's boundary is nowhere dense in the sense that its double negation is zero, and in our dual case, the π-coboundary is π-dense in the sense that its π-double-negation transform is $\mathbf{1}_U$:

$$\overset{\pi}{\neg}\overset{\pi}{\neg}\partial^\pi\sigma = \overset{\pi}{\neg}\overset{\pi}{\neg}\left(\sigma \vee \overset{\pi}{\neg}\sigma\right) = \mathbf{1}_U.$$

Lawvere defines the "core" of an element as its double negation and then proves that each element is equal to its boundary joined with its core. In the dual case of partitions, we have the π-version of that result.

Proposition 12 (Lawvere's boundary + core law for partitions)

$$\partial^\pi\sigma \wedge \overset{\pi}{\neg}\overset{\pi}{\neg}\sigma = \sigma \vee \pi.$$

Proof: This is easily proved from Ore's associability theorem using some basic identities such as the "law of contradiction" in $\boldsymbol{B}_\pi$: $\overset{\pi}{\neg}\sigma \wedge \overset{\pi}{\neg}\overset{\pi}{\neg}\sigma = \pi$ as well as $\sigma \preceq \overset{\pi}{\neg}\overset{\pi}{\neg}\sigma$ so that $\sigma \vee \overset{\pi}{\neg}\overset{\pi}{\neg}\sigma = \overset{\pi}{\neg}\overset{\pi}{\neg}\sigma$. Then using Ore's theorem:

$$\sigma \vee \pi = \sigma \vee \left(\overset{\pi}{\neg}\sigma \wedge \overset{\pi}{\neg}\overset{\pi}{\neg}\sigma\right) = \left(\sigma \vee \overset{\pi}{\neg}\sigma\right) \wedge \left(\sigma \vee \overset{\pi}{\neg}\overset{\pi}{\neg}\sigma\right) = \partial^\pi\sigma \wedge \overset{\pi}{\neg}\overset{\pi}{\neg}\sigma. \qquad \square$$

Restricting to $\sigma \in [\pi, 1]$, any σ that refines π (so that $\sigma \vee \pi = \sigma$) can be reconstructed from its π-double-negation $\overset{\pi}{\neg}\overset{\pi}{\neg}\sigma$ by taking the meet with its π-coboundary $\partial^\pi\sigma = \sigma \vee \overset{\pi}{\neg}\sigma$.

Corollary 5 *If $\sigma \in [\pi, 1]$, then $\sigma = \partial^\pi\sigma \wedge \overset{\pi}{\neg}\overset{\pi}{\neg}\sigma$.* $\qquad\square$

Lawvere showed that the Leibniz $(fg)' = f(g') + (f')g$ holds in the co-Heyting algebra of closed subsets by replacing the derivative by the boundary. In our case, the dual Leibniz rule holds using the π-coboundary.

Proposition 13 (co-Leibniz rule for partitions)

$$\partial^{\pi}(\sigma \vee \tau) = (\partial^{\pi}\sigma \vee \tau) \wedge (\sigma \vee \partial^{\pi}\tau).$$

Proof: Analyzing using B-slots, $\partial^{\pi}(\sigma \vee \tau) = (\sigma \vee \tau) \vee \overset{\pi}{\neg}(\sigma \vee \tau)$ so the blocks of $\sigma \vee \tau$ are the non-empty intersections, and the contents of the B-slot are $\mathbf{1}_B$ if there is a $C \in \sigma$ and $D \in \tau$ such that $B \subseteq C \cap D$, and otherwise are the restriction $(\sigma \vee \tau) \restriction B = \{C \cap D \cap B \neq \emptyset : C \in \sigma, D \in \tau\}$. On the RHS, the B-slot contents of $\partial^{\pi}\sigma \vee \tau = \sigma \vee \tau \vee \overset{\pi}{\neg}\sigma$ are $\mathbf{1}_B$ if there is a $C \in \sigma$ with $B \subseteq C$ and otherwise $(\sigma \vee \tau) \restriction B$. The B-slot contents of $\sigma \vee \partial^{\pi}\tau = \sigma \vee \tau \vee \overset{\pi}{\neg}\tau$ are $\mathbf{1}_B$ if there is a $D \in \tau$ with $B \subseteq D$ and otherwise $(\sigma \vee \tau) \restriction B$. Hence the B-slot contents of the meet $(\partial^{\pi}\sigma \vee \tau) \wedge (\sigma \vee \partial^{\pi}\tau)$ are $\mathbf{1}_B$ if there is both a $C \in \sigma$ and a $D \in \tau$ with $B \subseteq C, D$ and is otherwise $(\sigma \vee \tau) \restriction B$. Hence both sides have the same B-slot content and thus are equal. $\qquad\square$

2.5   Partition logic via the RST-closure space $\wp(U \times U)$

2.5.1   The RST-closure space $\wp\,(U \times U)$

In the Boolean logic of subsets, the standard semantics is the powerset Boolean algebra $\wp\,(U)$ and the analogous structure for partition logic would seem to be the algebra of partitions $\Pi\,(U)$. But partition logic is more complicated and has a richer theory. The partition algebra $\Pi\,(U)$ is isomorphic to the algebra of open subsets $\mathcal{O}\,(U \times U)$ where an open subset of $U \times U$ (a partition relation) is the complement of the Reflexive-Symmetric-Transitive or RST closure of any subset of $U \times U$. The powerset Boolean algebra on $U \times U$ (i.e., the BA of all binary relations on U) equipped with the RST-closure operation will be called the *RST-closure space of* U, $\wp\,(U \times U)$.

The RST-closure is a closure operation in the standard sense [10, p. 111]: for any $V, W \subseteq U \times U$,

- $V \subseteq \overline{V}$,

- $\overline{V} = \overline{\overline{V}}$ (idempotent) and

- If $V \subseteq W$, then $\overline{V} \subseteq \overline{W}$ (monotone).

It follows that the *closed subsets*, i.e., $V = \overline{V}$, form a complete lattice, the complete lattice of equivalence relations on U in our case. The complements of the closed subsets of $\wp(U \times U)$ are the open subsets which are the partition relations (ditsets of partitions) and give the isomorphism $\Pi(U) \cong \mathcal{O}(U \times U)$. Arbitrary intersections of closed subsets are closed and thus arbitrary unions of open subsets are open. But the RST-closure is not topological in that sense that a union of closed subsets is not necessarily closed, and thus an intersection of open subsets is not necessarily open. Just as every subset V has a closure (which is an equivalence relation, the intersection of all closed subsets containing V), so every subset has an interior $\text{int}(V) = \overline{(V^c)}^c$, the complement of the RST-closure of the complement of V, which is the ditset of a partition on U which might be denoted. $\lambda(\text{int}(V))$, read as "the partition whose ditset is $\text{int}(V)$."

The adjunctive definition of the partition implication:

$$\text{dit}(\tau) \cap \text{dit}(\sigma) \subseteq \text{dit}(\pi) \text{ iff } \text{dit}(\tau) \subseteq \text{int}(\text{dit}(\sigma)^c \cup \text{dit}(\pi))$$

was formulated in $\wp(U \times U)$. It hints that the deeper analysis of partition logic might be better formulated in that closure space rather than $\Pi(U)$.

By working in a powerset BA, we also have the luxury of using illustrations using Venn diagrams. Each partition π on U partitions the space into $\text{dit}(\pi)$ and $\text{indit}(\pi)$. The largest consequence of working with $\wp(U \times U)$ is the logical definition of information as distinctions so the information given in a partition π is given by its ditset $\text{dit}(\pi)$. For finite U, a probability measure on $U \times U$ is obtained by using the product measure from U where the probability measure on U could be equiprobable points or point probabilities. Then the probability assigned to $\text{dit}(\pi)$ is the logical entropy $h(\pi)$. The development of logical entropy and its relation to Shannon entropy provides new logical foundations for information theory that has been developed elsewhere [32]. Hence the focus here is on partition logic itself.

The "structure theorem" for those two complementary subsets gives:

$$\text{indit}(\pi) = \cup_{B \in \pi} B \times B \text{ and } \text{dit}(\pi) = \cup_{B, B' \in \pi, B \neq B'} B \times B'$$

where the union for $\text{dit}(\pi)$ are interpreted to include both $B \times B'$ and $B' \times B$. Given two partitions π and σ, the two partitions on $U \times U$ intersect to give $2^2 = 4$ atomic areas. The unions in the following propositions are interpreted as including the permutations of the factors in the Cartesian products.

Proposition 14 *Structure theorem for 4-part partition of closure space $U \times U$ given by any σ and π:*

1. $\displaystyle\bigcup_{B \in \pi, C \in \sigma} (B - (B \cap C)) \times (C - (B \cap C)) = \mathrm{dit}\,(\pi) \cap \mathrm{dit}\,(\sigma)\,;$

2. $\displaystyle\bigcup_{B \in \pi, C \in \sigma} (B - (B \cap C)) \times (B \cap C) = (\mathrm{indit}\,(\pi) \cap \mathrm{dit}\,(\sigma))\,;$

3. $\displaystyle\bigcup_{B \in \pi, C \in \sigma} (B \cap C) \times (C - (B \cap C)) = (\mathrm{indit}\,(\sigma) \cap \mathrm{dit}\,(\pi))\,;$ and

4. $\displaystyle\bigcup_{B \in \pi, C \in \sigma} (B \cap C) \times (B \cap C) = \mathrm{indit}\,(\sigma) \cap \mathrm{indit}\,(\pi).$

Proof: Part 1: The union is disjoint since each summand will differ by a B or a C. Assume $(u, u') \in (B - (B \cap C)) \times (C - (B \cap C))$ so $u \in B - (B \cap C) = B - C$ and $u' \in C - B$. Since $u' \in C - B$, it must be in a different block of π than B so $(u, u') \in \mathrm{dit}\,(\pi)$. Similarly since $u \in B - C$, it must be in a different block of σ than C so $(u, u') \in \mathrm{dit}\,(\sigma)$ and thus $(u, u') \in \mathrm{dit}\,(\pi) \cap \mathrm{dit}\,(\sigma)$. Conversely if $(u, u') \in \mathrm{dit}\,(\pi) \cap \mathrm{dit}\,(\sigma)$, then u is in some block $B \in \pi$ and u' is in some block $C \in \sigma$. But since (u, u') is a dit of both partitions u cannot be in C and u' cannot be in B so $u \in B - C$ and $u' \in C - B$ and thus $(u, u') \in (B - C) \times (C - B)$.

Part 2: Let $(u, u') \in (B - C) \times (B \cap C)$ so u is in B but not in C while u' is in both so $(u, u') \in \mathrm{indit}\,(\pi) \cap \mathrm{dit}\,(\sigma)$. Conversely, if $(u, u') \in \mathrm{indit}\,(\pi) \cap \mathrm{dit}\,(\sigma)$ then for $u, u' \in B \in \pi$ and $u' \in C \in \sigma$, then since $(u, u') \in \mathrm{dit}\,(\sigma)$, u must be in a different block $C' \in \sigma$ so $u \in B - C$ and $u' \in B \cap C$.

Parts 3 and 4 proved in a similar manner.   □

These structure theorems extend to the 2^n atomic areas determined by n partitions on U.

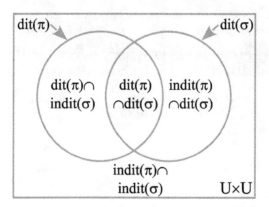

Figure 2.2: Four atomic areas in $U \times U$ Venn diagram for two partitions.

The structure theorem shows how ditsets or inditsets of a partition can be constructed from how the partition 'interacts' with another partition. For instance, the disjoint union of the two atomic areas in Figure 2.2 giving $\text{dit}(\pi)$ is:

$$\text{dit}(\pi) = (\text{dit}(\pi) \cap \text{dit}(\sigma)) \cup (\text{dit}(\pi) \cap \text{indit}(\sigma))$$
$$= \bigcup_{B \in \pi, C \in \sigma} (B - C) \times (C - B) \cup \bigcup_{B \in \pi, C \in \sigma} (B \cap C) \times (C - B).$$

In logical information theory (developed in a later chapter as an application of partition logic) for a finite U with the product probability distribution on $U \times U$, the logical entropy of a partition is the probability of its ditset, $h(\pi) = \Pr(\text{dit}(\pi))$. Then $\text{dit}(\pi) \cap \text{dit}(\sigma)$ represents the mutual information (i.e., common distinctions) of π and σ whose probability is $m(\pi, \sigma) = \Pr(\text{dit}(\pi) \cap \text{dit}(\sigma))$ and $\text{dit}(\pi) \cap \text{indit}(\sigma) = \text{dit}(\pi) - \text{dit}(\sigma)$ represents the information given by π after we take away the information of σ whose probability is $h(\pi|\sigma)$. Hence the representation of $\text{dit}(\pi)$ as the disjoint union gives the representation of the information in π as the sum of the information in common with σ plus the information in π after the information in σ has been taken away:

$$h(\pi) = m(\pi, \sigma) + h(\pi|\sigma).$$

The structure theorem also allows us to connect properties of partitions with properties of their ditsets. For instance, two partitions π and σ on a finite U are said to be (stochastically) *independent* if with equiprobable points, for all $B \in \pi$ and $C \in \sigma$, $p_{B \cap C} = p_B p_C$, i.e., $\frac{|B \cap C|}{|U|} = \frac{|B|}{|U|} \frac{|C|}{|U|}$ or $|B \cap C||U| = |B||C|$. The equiprobable distribution on U induces the equiprobable distribution on $U \times U$, so we also have the notion of two events $V, W \subseteq U \times U$ being *independent* if $p_{V \cap W} = p_V p_W$, i.e., $\frac{|V \cap W|}{|U \times U|} = \frac{|V|}{|U \times U|} \frac{|W|}{|U \times U|}$ or $|V \cap W||U|^2 = |V||W|$.

Proposition 15 *If π and σ are independent partitions, then $\text{dit}(\pi)$ and $\text{dit}(\sigma)$ are independent events in $U \times U$ (equiprobable points on finite U).*

Proof: If π and σ are independent partitions, then $|B \cap C||U| = |B||C|$ for all $B \in \pi$ and $C \in \sigma$. Since the unions in the Structure Theorem are disjoint, the unions can be expressed as sums. Thus from Part 1, we have:

$$|\text{dit}\,(\pi) \cap \text{dit}\,(\sigma)| = \sum_{B\in\pi,C\in\sigma} (|B| - |B \cap C|) \times (|C| - |B \cap C|)$$

$$|\text{dit}\,(\pi) \cap \text{dit}\,(\sigma)| = \sum_{B\in\pi,C\in\sigma} (|B| - |B \cap C|)\,(|C| - |B \cap C|)$$

$$= \sum_{B\in\pi,C\in\sigma} \left(|B| - \frac{|B|\,|C|}{|U|}\right)\left(|C| - \frac{|B|\,|C|}{|U|}\right)$$

$$= \frac{1}{|U|^2} \sum_{B\in\pi,C\in\sigma} (|B|\,|U| - |B|\,|C|)\,(|C|\,|U| - |B|\,|C|)$$

$$= \frac{1}{|U|^2} \sum_{B\in\pi,C\in\sigma} (|B|\,(|U| - |C|))\,(|C|\,(|U| - |B|))$$

$$= \frac{1}{|U|^2} \left[\sum_{B\in\pi} |B|\,(|U| - |B|)\right]\left[\sum_{C\in\sigma} |C|\,(|U| - |C|)\right]$$

$$= \frac{1}{|U|^2} |\text{dit}\,(\pi)|\,|\text{dit}\,(\sigma)|\,.\square$$

Furthermore, dividing both sides of the equation by $|U|^2$ gives for independence in the equiprobable case;

$$m\,(\pi,\sigma) = h\,(\pi)\,h\,(\sigma)$$

as one would expect since logical entropies are the two-draws-from-U probabilities of getting a distinction. Hence in the case of independence, the probability that two draws yield a distinction of both π and σ is the product of the probability of getting a distinction of π times the probability of getting a distinction of σ.

For an example, consider the independent partitions $\pi = \{\{a,b\},\{c,d\}\}$ and $\sigma = \{\{a,c\},\{b,d\}\}$. The ditset of π is $\{(a,c),(a,d),(b,c),(b,d),...\}$ where the ellipsis stands for the opposite ordered pairs, and $\text{dit}\,(\sigma) = \{(a,b),(a,d),(c,b),(c,d),...\}$ so $|\text{dit}\,(\pi) \cap \text{dit}\,(\sigma)| = |\{(a,d),(b,c),...\}| = 4$. And $\frac{1}{|U|^2}|\text{dit}\,(\pi)|\,|\text{dit}\,(\sigma)| = \frac{1}{16}8 \times 8 = 4.\checkmark$ Dividing through by $|U|^2$ gives:

$$m\,(\pi,\sigma) = \frac{|\text{dit}(\pi)\cap\text{dit}(\sigma)|}{|U|^2} = \frac{1}{4} \text{ and } h\,(\pi)\,h\,(\sigma) = \frac{|\text{dit}(\pi)|}{|U|^2}\frac{|\text{dit}(\sigma)|}{|U|^2} = \frac{8}{16}\frac{8}{16} = \frac{1}{4}.\checkmark$$

2.5.2   The sixteen binary logical operations

Starting with two subsets $\text{dit}\,(\sigma)$ and $\text{dit}\,(\tau)$ in a Venn diagram, there are sixteen possible unions of the four atomic areas (each union defined by whether each

atomic area is included or not), so there are sixteen definable areas in the
Venn diagram. The interiors of the sixteen definable areas define the ditsets
of the sixteen binary logical operations on partitions–and their closures define
the sixteen binary logical operations on equivalence relations. The operations
have a natural pairing since each definable area has a definable complement
so the interiors of those complementary subsets gives a complement-pairing of
operations. For instance, the interior int [dit $(\sigma) \cap$ dit (τ)] is dit $(\sigma \wedge \tau)$ and the
interior of its complement int [indit $(\sigma) \cup$ indit (τ)] is dit $(\sigma|\tau)$. Previously, we
defined that two partitions φ and φ' are orthogonal if $\neg\varphi \vee \neg\varphi' = 1_U$. The
negation $\neg\varphi = \varphi \Rightarrow 0_U$ is either 0_U if $\varphi \neq 0_U$ or 1_U if $\varphi = 0_U$. Hence φ and
φ' are orthogonal iff one or both are 0_U. There is the partition identity:

$$\neg\varphi \vee \neg\varphi' = \varphi \Rightarrow \neg\varphi' = \varphi' \Rightarrow \neg\varphi$$

which carries the sense that when $\neg\varphi \vee \neg\varphi' = 1_U$ then if one is not 0_U, then
the other has to be 0_U. Since the interior of a definable area and the interior
of its complement are disjoint, we know from the Common Dits Theorem that
one of them has to be 0_U so the complementary-dual partitions are orthogonal.
Table 2.4 gives the complementary duality pairings.

Defined subset $S \subseteq U \times U$	int (S) is dit set	int (S^c) is dit set
$U \times U$	1_U	0_U
dit $(\sigma) \cup$ dit (τ)	$\sigma \vee \tau$	$\sigma\overline{\vee}\tau = \neg\sigma \wedge \neg\tau$
indit $(\sigma) \cup$ dit (τ)	$\sigma \Rightarrow \tau$	$\tau \nLeftarrow \sigma = \sigma \wedge \neg\tau$
dit (τ)	τ	$\neg\tau$
dit $(\sigma) \cup$ indit (τ)	$\tau \Rightarrow \sigma$	$\sigma \nLeftarrow \tau = \tau \wedge \neg\sigma$
dit (σ)	σ	$\neg\sigma$
(dit $(\sigma) \cap$ dit $(\tau)) \cup$ (indit $(\sigma) \cap$ indit $(\tau))$	$(\sigma \Rightarrow \tau) \wedge (\tau \Rightarrow \sigma)$	$(\sigma \vee \tau) \wedge (\sigma|\tau)$
dit $(\sigma) \cap$ dit (τ)	$\sigma \wedge \tau$	$\sigma|\tau$

Table 2.4: Complementary-dual pairing of sixteen binary logical operations.

In the standard Venn diagram for two subsets dit (σ) and dit (τ), there are
four atomic areas and $2^4 = 16$ definable subsets. Those subsets can be arranged
as the vertices of the 4-hypercube graph Q_4 where the edges are the Hasse
diagram inclusions, i.e., an edge indicates an inclusion with no other subsets in
between. Since the interior operator is monotonic, we can take the interiors of
all the subsets so the vertices can be labeled with the corresponding partitions
and the edges indicate refinement relations. In Figured 2.3, the equivalence
$(\sigma \Rightarrow \tau) \wedge (\tau \Rightarrow \sigma)$ and the inequivalence (or symmetric difference) have been
abbreviated respectively as $\sigma \Longleftrightarrow \tau$ and $\sigma\Delta\tau$.

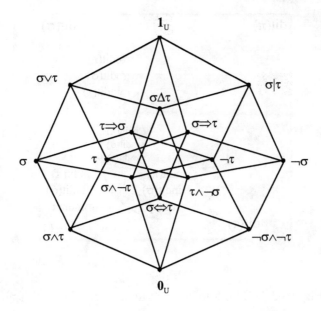

Figure 2.3: Hypercube graph representation of the sixteen binary logical operations on partitions.

With each vertex in the graph, there is an opposite vertex (representing the complementary subset) and the corresponding partitions are the complementary duals. For instance, the dual of $\sigma \vee \tau$ is $\neg\sigma \wedge \neg\tau$ and the dual of $\sigma\Delta\tau$ is $\sigma \Longleftrightarrow \tau$. Furthermore, it might be noticed that whenever two partitions have a common immediate predecessor, then since $\text{int}(S \cap T) = \text{int}(\text{int}(S) \cap \text{int}(T))$, then that common predecessor is their meet. For instance, $(\sigma \vee \tau) \wedge (\sigma|\tau) = \sigma\Delta\tau$ and $\sigma \wedge (\sigma\Delta\tau) = \sigma \wedge \neg\tau$. The same relation for common successors does not hold for the join (unless we dualize to equivalence relations). For instance, $\sigma|\tau$ is an immediate common successor of $\neg\sigma$ and $\neg\tau$ but is not necessarily their join.

Due to the rather 'severe' nature of the $\mathbf{0}_U$-negation $\sigma \Rightarrow \mathbf{0}_U$, we have seen that more interesting structure emerges by relativizing the operations to some fixed π. In the usual Venn diagram for three areas, $\text{dit}(\pi)$, $\text{dit}(\sigma)$, and $\text{dit}(\tau)$, there are eight atomic areas. But to work in the segment $[\pi, \mathbf{1}_U]$, we focus on the areas that include $\text{dit}(\pi)$. There are four atomic areas outside of $\text{dit}(\pi)$ as in Figure 2.4 that could be unioned with $\text{dit}(\pi)$.

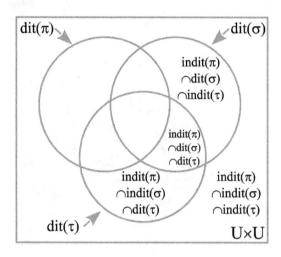

Figure 2.4: The four atomic areas outside of dit (π).

But since we are working in the BA $\wp(U \times U)$, we can distribute the union of dit (π) with these four areas to get the four π-relativized 'atomic' areas:

- dit $(\pi) \cup ($indit $(\pi) \cap$ dit $(\sigma) \cap$ indit $(\tau)) = $ dit $(\pi) \cup ($dit $(\sigma) \cap$ indit $(\tau))$;

- dit $(\pi) \cup ($indit $(\pi) \cap$ dit $(\sigma) \cap$ dit $(\tau)) = $ dit $(\pi) \cup ($dit $(\sigma) \cap$ dit $(\tau))$;

- dit $(\pi) \cup ($indit $(\pi) \cap$ indit $(\sigma) \cap$ dit $(\tau)) = $ dit $(\pi) \cup ($indit $(\sigma) \cap$ dit $(\tau))$; and

- dit $(\pi) \cup ($indit $(\pi) \cap$ indit $(\sigma) \cap$ indit $(\tau)) = $ dit $(\pi) \cup ($indit $(\sigma) \cap$ indit $(\tau))$.

There again sixteen definable unions of these four areas and the interiors of those sixteen areas give the sixteen π-relativized 'binary' logical operations on partition–which are actually ternary operations of a special form. Moreover, each of the areas has a complement in $U \times U$ with dit (π) added back in to give the π-complementary-dual π-relativized partition operations in Table 2.5. In Table 2.5, $S = $ dit (σ), $T = $ dit (τ), $P = $ dit (π), and ^{c} is complement.

$V \subseteq U \times U$	int $(V \cup P)$ is ditset	int $(V^c \cup P)$ is ditset
$V = U \times U$	$\mathbf{1}_U$	π
$V = S \cup T$	$\sigma \vee \tau \vee \pi$	$\overset{\pi}{\neg}\sigma \wedge \overset{\pi}{\neg}\tau$
$V = S^c \cup T$	$\sigma \Rightarrow (\tau \vee \pi)$	$(\sigma \vee \pi) \wedge \overset{\pi}{\neg}\tau$
$V = T$	$\tau \vee \pi$	$\overset{\pi}{\neg}\tau$
$V = S \cup T^c$	$\tau \Rightarrow (\sigma \vee \pi)$	$\overset{\pi}{\neg}\sigma \wedge (\tau \vee \pi)$
$V = S$	$\sigma \vee \pi$	$\overset{\pi}{\neg}\sigma$
$V = (S \cap T) \cup (S^c \cap T^c)$	$(\sigma \Rightarrow (\tau \vee \pi)) \wedge (\tau \Rightarrow (\sigma \vee \pi))$	$(\sigma \vee \tau \vee \pi) \wedge (\sigma|_\pi \tau)$
$V = S \cap T$	$(\sigma \vee \pi) \wedge (\tau \vee \pi)$	$\sigma|_\pi \tau$

Table 2.5: π-complementary-duals of π-relativized partitions.

Figure 2.5 is the hypercube graph drawn by replacing each partition in Figure 2.3 by its π-relativized version and the same relationships hold.

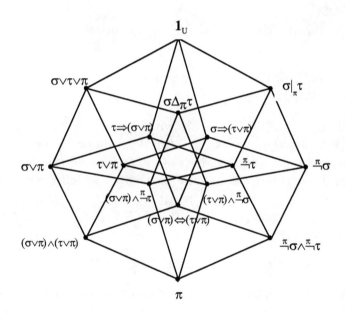

Figure 2.5: Hypercube graph for sixteen π-relativized binary logical operations.

It is then immediate that the meet of each partition with its π-complementary dual is π as indicated in Table 2.5.

Meet of π-complementary-duals $= \pi$
$\mathbf{1}_U \wedge \pi = \pi$
$(\sigma \vee \tau \vee \pi) \wedge \left(\overset{\pi}{\neg}\sigma \wedge \overset{\pi}{\neg}\tau \right) = \pi$
$(\sigma \Rightarrow (\tau \vee \pi)) \wedge \left((\sigma \vee \pi) \wedge \overset{\pi}{\neg}\tau \right) = \pi$
$(\tau \vee \pi) \wedge \left(\overset{\pi}{\neg}\tau \right) = \pi$
$(\tau \Rightarrow (\sigma \vee \pi)) \wedge \left(\overset{\pi}{\neg}\sigma \wedge (\tau \vee \pi) \right) = \pi$
$(\sigma \vee \pi) \wedge \left(\overset{\pi}{\neg}\sigma \right) = \pi$
$(\sigma \Rightarrow (\tau \vee \pi)) \wedge (\tau \Rightarrow (\sigma \vee \pi)) \wedge (\sigma \vee \tau \vee \pi) \wedge (\sigma|_\pi \tau) = \pi$
$((\sigma \vee \pi) \wedge (\tau \vee \pi)) \wedge (\sigma|_\pi \tau) = \pi$

Table 2.5: Meet of π-complementary dual partitions.

One immediate result is that the π-negation of the LHS formulas in Table 2.5 are $\pi \Rightarrow \pi = \mathbf{1}_U$, so the π-negation of all the LHS formulas are partition tautologies, e.g., $\overset{\pi}{\neg} \left[(\sigma \vee \pi) \wedge \left(\overset{\pi}{\neg}\sigma \right) \right] = \mathbf{1}_U$.

This basic theorem is essentially a π-generalization of the Common-Dits Theorem.

Theorem 11 *For any $V \subseteq U \times U$, each block $B \in \pi$ is contained in a block of* int $(V \cup \mathrm{dit}\,(\pi))$ *or in a block of* int $(V^c \cup \mathrm{dit}\,(\pi))$.

Proof: Taking σ as the partition with the ditset int $(V \cup \mathrm{dit}\,(\pi))$, i.e., $\sigma = \lambda\,(\mathrm{int}\,(V \cup \mathrm{dit}\,(\pi)))$, and $\tau = \lambda\,(\mathrm{int}\,(V^c \cup \mathrm{dit}\,(\pi)))$, we have $\mathrm{dit}\,(\pi) \subseteq \mathrm{dit}\,(\sigma) \subseteq V \cup \mathrm{dit}\,(\pi)$ and $\mathrm{dit}\,(\pi) \subseteq \mathrm{dit}\,(\tau) \subseteq V^c \cup \mathrm{dit}\,(\pi)$ so that $\mathrm{dit}\,(\sigma) \cap \mathrm{dit}\,(\tau) = \mathrm{dit}\,(\pi)$. Complementing gives $\mathrm{indit}\,(\sigma) \cup \mathrm{indit}\,(\tau) = \mathrm{indit}\,(\pi)$. Then for any $B \in \pi$, consider the two cut-down partitions on B, $\sigma \restriction B$ and $\tau \restriction B$. Then $(B \times B) \cap (\mathrm{indit}\,(\sigma) \cup \mathrm{indit}\,(\tau)) = B \times B$ and $(B \times B) \cap \mathrm{indit}\,(\sigma) = \mathrm{indit}\,(\sigma \restriction B)$ and $(B \times B) \cap \mathrm{indit}\,(\tau) = \mathrm{indit}\,(\tau \restriction B)$. Thus

$$\mathrm{indit}\,(\tau \restriction B) \cup \mathrm{indit}\,(\tau \restriction B) = B \times B$$

so by the equivalence relation version of the Common-Dits Theorem, either $\mathrm{indit}\,(\sigma \restriction B) = B \times B$ or $\mathrm{indit}\,(\tau \restriction B) = B \times B$ so $B \subseteq C$ for some $C \in \sigma$ or $B \subseteq D$ for some $D \in \tau$. □

Of course, a block $B \in \pi$ is contained in any π-regular partition $\overset{\pi}{\neg}\sigma$ or its π-negation $\overset{\pi}{\neg}\overset{\pi}{\neg}\sigma$ in B_π but the theorem is more general for any φ (taking $V = \mathrm{dit}\,(\varphi)$) and its π-complementary-dual $\lambda\,(\mathrm{int}\,(\mathrm{indit}\,(\varphi) \cup \mathrm{dit}\,(\pi)))$.

Corollary 6 *For any $V \in U \times U$, for $\varphi = \lambda\,(\mathrm{int}\,(V \cup \mathrm{dit}\,(\pi)))$ and $\varphi' = \lambda\,(\mathrm{int}\,(V^c \cup \mathrm{dit}\,(\pi)))$, φ and φ' are π-orthogonal.*

Proof: Now φ and φ' are π-orthogonal if $\overset{\pi}{\neg}\varphi \vee \overset{\pi}{\neg}\varphi' = \mathbf{1}_U$. Since B is contained in a block of φ or φ', the B-slot in $\overset{\pi}{\neg}\varphi$ or $\overset{\pi}{\neg}\varphi'$ will be $\mathbf{1}_B$, so the join is $\mathbf{1}_U$. $\square$

Corollary 7 *The pairs of π-complementary dual operations on any σ and τ are π-orthogonal.* $\square$

It should be noted that for different V's, $\lambda(\text{int}(V \cup \text{dit}(\pi)))$ can be the same partition with different π-orthogonals $\lambda(\text{int}(V^c \cup \text{dit}(\pi)))$. For instance with $\pi = \mathbf{0}_U$, for $V = \text{dit}(\sigma) \cap \text{dit}(\tau)$, $\sigma \wedge \tau = \lambda(\text{int}(V))$ and for $V' = \text{int}[\text{dit}(\sigma) \cap \text{dit}(\tau)] = \text{dit}(\sigma \wedge \tau)$, then we also have $\sigma \wedge \tau = \lambda(\text{int}(V'))$. But taking complements,

$$\lambda(\text{int}(V^c)) = \lambda(\text{int}(\text{indit}(\sigma) \cup \text{indit}(\tau))) = \sigma | \tau \text{ and}$$
$$\lambda(\text{int}(V'^c)) = \lambda(\text{int}(\text{indit}(\sigma \wedge \tau))) = \neg(\sigma \wedge \tau) = (\sigma \wedge \tau) \Rightarrow \mathbf{0}_U$$

which are not necessarily the same, although there is the ordering relationship:

$$\neg\sigma \vee \neg\tau \precsim \sigma | \tau \precsim \neg(\sigma \wedge \tau).$$

2.5.3 The π-orthogonal algebra $\mathcal{A}_\pi = [\pi, \mathbf{1}_U]$

We have seen a number of suggestive analogies between the upper segment $[\pi, \mathbf{1}_U]$ of $\Pi(U)$ and the Heyting algebra of open subsets of a topological space. All the sixteen binary logical operations on partitions can be defined relative to π, and in particular, negation is then π-negation. In a Heyting algebra and in the upper segment $[\pi, \mathbf{1}_U]$, the negated elements form a Boolean algebra. In the Heyting algebra of open subsets, there is a natural notion of disjointness, namely the intersection of open subsets is the null set, and the negation of an open subset is the largest open subset disjoint from it. Similarly, in the upper segment $[\pi, \mathbf{1}_U]$, there is the notion of π-orthogonality between partitions and the π-negation of a partition is the largest (most refined) partition π-orthogonal to it. All of these similarities suggest that the upper segment $[\pi, \mathbf{1}_U]$ is an algebraic object worthy of study in its own right. Roughly speaking, it is to partitions (i.e., open subsets of $\wp(U \times U)$) what the Heyting algebra is to open subsets of a topological space. Hence we will call the upper segments $[\pi, \mathbf{1}_U]$, π *-orthogonal algebras*, and denote them as $\mathcal{A}_\pi$.

The π-orthogonal algebras are of independent interest since they are not Heyting algebras as we have seen for a number of reasons such as they are not distributive as lattices and the π-negation is not a (relative) pseudo-complementation. Moreover, there is not inclusion either way between partition tautologies and Heyting algebra validities. However one way to study the π-orthogonal algebras is to look at the similarities and dissimilarities with Heyting algebras of open subsets. Hence we start with a number of corollaries of previous results that show similarities.

Corollary 8 *If φ and φ' are π-orthogonal, then $\varphi \vee \pi$ and $\varphi' \vee \pi$ are π-orthogonal and $(\varphi \vee \pi) \wedge (\varphi' \vee \pi) = \pi$. In particular, if $\pi \precsim \varphi, \varphi'$, then $\varphi \wedge \varphi' = \pi$.*

Proof: Since $\varphi \vee \pi$ and $\varphi' \vee \pi$ refine π and each B must be contained in a block of φ or φ', B must equal a block of $\varphi \vee \pi$ or $\varphi' \vee \pi$, so the B-slot in the join is always $\mathbf{1}_B$ and in the meet is $\mathbf{0}_B = B$, i.e., $(\varphi \vee \pi) \wedge (\varphi' \vee \pi) = \pi$. $\square$

For an example using the partition notation that concatenated letters (with no commas in between) are in the same block, let $\varphi = \{abc, def\}$, $\varphi' = \{abe, cdf\}$ and $\pi = \{ab, cd, ef\}$. Then $\overset{\pi}{\neg}\varphi = \{a, b, cd, e, f\}$ and $\overset{\pi}{\neg}\varphi' = \{a, b, c, d, ef\}$ so φ and φ' are π-orthogonal but $\varphi \wedge \varphi' = \mathbf{0}_U$. But $\varphi \vee \pi = \{ab, c, d, ef\}$ and $\varphi' \vee \pi = \{ab, cd, e, f\}$, so they are still π-orthogonal and $(\varphi \vee \pi) \wedge (\varphi' \vee \pi) = \pi$.

Corollary 9 *For any partition φ, $\overset{\pi}{\neg}\varphi$ is the maximal (i.e., most refined) π-orthogonal partition to φ.*

Proof: Any φ and $\overset{\pi}{\neg}\varphi$ are π-orthogonal since $\overset{\pi}{\neg}\varphi \vee \overset{\pi}{\neg}\overset{\pi}{\neg}\varphi = \mathbf{1}_U$, i.e., the law of excluded middle in the BA $\mathcal{B}[\pi, \mathbf{1}_U]$. If φ' is any other π-orthogonal partition to φ, then $\varphi' \Rightarrow \overset{\pi}{\neg}\varphi = \mathbf{1}_U$ so $\varphi' \precsim \overset{\pi}{\neg}\varphi$. $\square$

It is tempting to think that $\varphi \wedge \varphi' = \pi$ is equivalent to π-orthogonality in analogy with disjointness in a Heyting algebra. But that is not so since for $\varphi = \{\{a, b\}, \{c\}\}$, $\varphi' = \{\{a\}, \{b, c\}\}$, and $\pi = \mathbf{0}_U$, then $\varphi \wedge \varphi' = \mathbf{0}_U$ but φ and φ' are not orthogonal. The meet-related characterization of π-orthogonality is just a little more subtle.

Corollary 10 *If $\pi \precsim \sigma, \tau$ and σ and τ are π-orthogonal, then σ and τ are associable.*

Proof: Since $\pi \precsim \sigma, \tau$, for any $B \in \pi$, then exactly B must be a block in either σ or τ so each block in the meet $\sigma \wedge \tau = \pi$ must be a block of σ or τ, so they are associable. $\square$

The converse does not hold since each partition is trivially associable with itself.

Corollary 11 *For $\pi \precsim \sigma, \tau$; σ and τ are π-orthogonal iff dit $(\sigma) \cap$ dit $(\tau) =$ dit (π).*

Proof: If $\pi \precsim \sigma, \tau$ and σ and τ are π-orthogonal, then σ and τ are associable so by a previous lemma, dit $(\sigma) \cap$ dit $(\tau) = $ dit $(\sigma \wedge \tau) = $ dit (π). Conversely, if dit $(\sigma) \cap$ dit $(\tau) = $ dit (π), then indit $(\sigma) \cup$ indit $(\tau) = $ indit (π). Thus indit $(\sigma) \cup$ indit $(\tau) \cup$ dit $(\pi) = U \times U$ so indit $(\tau)^c = $ dit $(\tau) \subseteq$ indit $(\sigma) \cup$ dit (π) and thus taking interiors,

$$\text{int } [\text{dit } (\tau)] = \text{dit } (\tau) \subseteq \text{int } [\text{indit } (\sigma) \cup \text{dit } (\pi)] = \text{dit } (\sigma \Rightarrow \pi)$$

so $\tau \precsim \overset{\pi}{\neg}\sigma$. Then $\overset{\pi}{\neg}\overset{\pi}{\neg}\sigma \precsim \overset{\pi}{\neg}\tau$ and

$$1_U = \overset{\pi}{\neg}\sigma \vee \overset{\pi}{\neg}\overset{\pi}{\neg}\sigma \precsim \overset{\pi}{\neg}\sigma \vee \overset{\pi}{\neg}\tau$$

so $1_U = \overset{\pi}{\neg}\sigma \vee \overset{\pi}{\neg}\tau$, i.e., σ and τ are π-orthogonal. $\square$

Lemma 12 *If $\overset{\pi}{\neg}\sigma \preceq \psi$, then $\psi = \overset{\pi}{\neg}\sigma \vee \left(\psi \wedge \overset{\pi}{\neg}\sigma'\right)$ for any π-orthogonal σ'.*

Proof: If $\exists C \in \sigma$ with $B \subseteq C$, then $\left(\overset{\pi}{\neg}\sigma\right)_B = 1_B$ and thus $\psi_B = 1_B$. Otherwise, $\overset{\pi}{\neg}\sigma' = 1_B$ and $\left(\psi \wedge \overset{\pi}{\neg}\sigma'\right)_B = \psi_B$ so for each $B \in \pi$, the expression gives ψ_B. $\square$

Lemma 13 *For any $\pi \preceq \psi$, and any π-orthogonal σ, τ, $\psi = \left(\psi \wedge \overset{\pi}{\neg}\sigma\right) \vee \left(\psi \wedge \overset{\pi}{\neg}\tau\right)$.*

Proof: Since $\overset{\pi}{\neg}\sigma \vee \overset{\pi}{\neg}\tau = 1_U$, the B-slot is 1_B in one or the other, and for that one, say, $\left(\psi \wedge \overset{\pi}{\neg}\sigma\right)_B = \psi_B$. The other is either also ψ_B or B, and in either case, the join gives ψ_B. $\square$

Lemma 14 *For any $\pi \preceq \psi$, and any σ, $\psi = \left(\psi \vee \overset{\pi}{\neg}\sigma\right) \wedge \left(\psi \vee \overset{\pi}{\neg}\overset{\pi}{\neg}\sigma\right)$.*

Proof: By Ore's theorem, $\psi = \psi \vee \pi = \psi \vee \left(\overset{\pi}{\neg}\sigma \wedge \overset{\pi}{\neg}\overset{\pi}{\neg}\sigma\right) = \left(\psi \vee \overset{\pi}{\neg}\sigma\right) \wedge \left(\psi \vee \overset{\pi}{\neg}\overset{\pi}{\neg}\sigma\right)$. $\square$

Lemma 15 *If σ, σ' are π-orthogonal and τ, τ' are also π-orthogonal and all refine π, then $\sigma \vee \tau$ and $\sigma' \wedge \tau'$ are π-orthogonal.*

Proof: We have $\overset{\pi}{\neg}\sigma \vee \overset{\pi}{\neg}\sigma' = 1_U$ and the same for τ and τ'. What we need to show is that $\overset{\pi}{\neg}(\sigma \vee \tau) \vee \overset{\pi}{\neg}(\sigma' \wedge \tau') = 1_U$. Thus we assume that B is not discretized in $\overset{\pi}{\neg}(\sigma \vee \tau)$ which means that B was further distinguished or broken up in σ or in τ. But that means that there is a $C' \in \sigma'$ or a $D' \in \tau'$ such that $C' = B = D'$. Since both σ' and τ' refine π, the block B would be reconstructed whole in $\sigma' \wedge \tau'$ and thus B would be discretized in $\overset{\pi}{\neg}(\sigma' \wedge \tau')$. $\square$

Lemma 16 *If $\pi \preceq \sigma, \tau, \tau'$ and τ, τ' are π-orthogonal, then $\sigma \Rightarrow \tau$ and $\sigma \wedge \tau'$ are π-orthogonal.*

Proof: We have $\overset{\pi}{\neg}\tau \vee \overset{\pi}{\neg}\tau' = \mathbf{1}_U$ and $\pi \preceq \sigma$ which implies that $\pi \preceq (\sigma \Rightarrow \tau), (\sigma \wedge \tau')$. We need to show $\overset{\pi}{\neg}(\sigma \Rightarrow \tau) \vee \overset{\pi}{\neg}(\sigma \wedge \tau') = 1$. Thus we assume that $B \in \pi$ is not discretized in $\overset{\pi}{\neg}(\sigma \Rightarrow \tau)$ which means either 1) $D \in \tau$ was distinguished further from B so B is a block in τ' and is thus discretized in $\overset{\pi}{\neg}(\sigma \wedge \tau')$, or 2) that $D = B$ but was discretized in $\sigma \Rightarrow \tau$ which means that C was also equal to B. But then $C = B$ means that B survives as a block in $\sigma \wedge \tau'$ and thus B is discretized in $\overset{\pi}{\neg}(\sigma \wedge \tau')$. $\square$

Lemma 17 *If $\pi \preceq \varphi, \varphi'$ are π-orthogonal, i.e., $\overset{\pi}{\neg}\varphi \vee \overset{\pi}{\neg}\varphi' = \mathbf{1}_U$, then for any $\psi \in [\pi, \varphi]$ and $\psi' \in [\pi, \varphi']$, ψ, ψ' are also π-orthogonal.*

Proof: Since $\psi \preceq \varphi$, $\overset{\pi}{\neg}\varphi \preceq \overset{\pi}{\neg}\psi$ and similarly for ψ' so $\mathbf{1}_U = \overset{\pi}{\neg}\varphi \vee \overset{\pi}{\neg}\varphi' \preceq \overset{\pi}{\neg}\psi \vee \overset{\pi}{\neg}\psi'$. $\square$

Corollary 12 *Given π-orthogonal $\pi \precsim \varphi, \varphi'$ and any $\pi \preceq \psi$, $\varphi \wedge \psi$ and $\varphi' \wedge \psi$ are π-orthogonal.* $\square$

Corollary 13 *If φ, φ' are π-orthogonal, then $\varphi \Rightarrow \overset{\pi}{\neg}\varphi'$ and $\varphi' \Rightarrow \overset{\pi}{\neg}\varphi$ are partition tautologies.*

Proof: Being π-orthogonal means $\overset{\pi}{\neg}\varphi \vee \overset{\pi}{\neg}\varphi' = \mathbf{1}_U$ which in turn means that any $B \in \pi$ is contained in a block of φ or a block of φ'. If B is contained in a block of φ', then the B-slot in $\overset{\pi}{\neg}\varphi'$ is $\mathbf{1}_B$ and thus also in $\varphi \Rightarrow \overset{\pi}{\neg}\varphi'$. If B was not contained in a block of φ', then the B-slot in $\overset{\pi}{\neg}\varphi'$ is $\mathbf{0}_B = B$ but B then has to be contained in a block of φ so the B-slot in $\varphi \Rightarrow \overset{\pi}{\neg}\varphi'$ is again $\mathbf{1}_B$. Thus $\varphi \Rightarrow \overset{\pi}{\neg}\varphi'$ is a partition tautology and similarly for $\varphi' \Rightarrow \overset{\pi}{\neg}\varphi$ by symmetry. $\square$

Corollary 14 *If φ and φ' are π-orthogonal and φ is a partition tautology, then so is $\overset{\pi}{\neg}\varphi'$.*

Proof: Modus ponens $\left[\varphi \wedge \left(\varphi \Rightarrow \overset{\pi}{\neg}\varphi'\right)\right] \Rightarrow \overset{\pi}{\neg}\varphi'$ is a partition tautology so if φ is, then so is $\overset{\pi}{\neg}\varphi'$. $\square$

The proof technique of using the B-component or B-slot of a partition $\sigma \in [\pi, \mathbf{1}_U]$ is based on the representation of that upper segment as the product of the partition algebras $\Pi(B)$ for non-singleton $B \in \pi$:

$$[\pi, \mathbf{1}_U] \cong \prod \{\Pi(B) : B \in \pi, B \text{ non-singleton}\}\ [49, \text{p. } 192].$$

Each $\sigma \in [\pi, \mathbf{1}_U]$ corresponds to the product of the restrictions $\sigma \restriction B \in \Pi(B)$, and given an element of that product, σ is reconstructed by taking the union of all blocks in the chosen partitions in $\Pi(B)$ plus the singleton blocks. The π-regular partitions are the ones constructed with only $\mathbf{0}_B$ or $\mathbf{1}_B$ from each $\Pi(B)$, which make up the Boolean core $\mathcal{B}[\pi, \mathbf{1}_U]$ of the segment $[\pi, \mathbf{1}_U]$. The π-orthogonal pairs $\sigma, \tau \in [\pi, \mathbf{1}_U]$ are the pairs where for every (non-singleton) $B \in \pi$, $\sigma \restriction B = \mathbf{0}_B$ or $\tau \restriction B = \mathbf{0}_B$. The construction of $\overset{\pi}{\neg}\sigma$ means replacing each $\sigma \restriction B = \mathbf{0}_B$ by $\mathbf{1}_B$ and each $\sigma \restriction B \neq \mathbf{0}_B$ by $\mathbf{0}_B$.

Before turning to a syntactic approach, we list a number of partition tautologies in no particular order in the following Table 2.6. The bi-implication $\pi \Leftrightarrow \sigma$ is short for $(\pi \Rightarrow \sigma) \wedge (\sigma \Rightarrow \pi)$.

$\mathbf{1}_U$	$((\pi \wedge \sigma) \vee \pi) \Leftrightarrow (\pi \wedge (\pi \vee \sigma))$
$\sigma \Rightarrow (\pi \vee \sigma)$	$((\pi \vee \sigma) \wedge \pi) \Leftrightarrow (\pi \vee (\pi \wedge \sigma))$
$\pi \Rightarrow (\sigma \Rightarrow \pi)$	$(\sigma \Rightarrow (\tau \Rightarrow \pi)) \Rightarrow ((\sigma \wedge \tau) \Rightarrow \pi)$
$(\sigma \Rightarrow \pi) \Rightarrow \sigma) \Rightarrow (\pi \Rightarrow \sigma)$	$\sigma \vee \tau \Rightarrow ((\sigma \Rightarrow \pi) \Rightarrow ((\tau \Rightarrow \pi) \Rightarrow \pi))$
$(\pi \wedge \sigma) \Rightarrow \pi$	$(\sigma \wedge (\tau \Rightarrow \pi)) \Rightarrow ((\sigma \Rightarrow \tau) \Rightarrow \pi)$
$\pi \Rightarrow (\pi \wedge \pi)$	$(\tau \Rightarrow (\sigma \Rightarrow \pi)) \Leftrightarrow (\sigma \Rightarrow (\tau \Rightarrow \pi))$
$\sigma \Leftrightarrow (\sigma \wedge (\pi \vee \sigma))$	$(\sigma \Rightarrow (\sigma \Rightarrow \pi)) \Leftrightarrow \sigma \Rightarrow \pi$
$(\sigma \vee (\pi \wedge \sigma)) \Leftrightarrow \sigma$	$(\sigma \wedge (\sigma \Rightarrow \pi)) \Leftrightarrow (\sigma \wedge \pi)$
$((\pi \vee \sigma) \Rightarrow \pi) \Rightarrow (\sigma \Rightarrow \pi)$	$(\sigma \Rightarrow \pi) \Rightarrow ((\sigma \vee \tau) \Rightarrow (\pi \vee \tau))$
$(\sigma \wedge (\sigma \Rightarrow \pi)) \Rightarrow \pi$	$(\sigma \Rightarrow \pi) \Rightarrow ((\tau \Rightarrow \phi) \Rightarrow ((\sigma \vee \tau) \Rightarrow (\pi \vee \phi)))$
$\pi \vee \mathbf{1}_U$	$(\sigma \vee \tau) \Rightarrow ([(\sigma \Rightarrow \pi) \wedge (\tau \Rightarrow \pi)] \Rightarrow \pi)$

Table 2.6: Some partition tautologies not involving negation

The following Table 2.7 lists a number of partition tautologies involving π-negation for any π.

$\sigma \Rightarrow \overset{\pi}{\neg}\overset{\pi}{\neg}\sigma$	$(\tau \Rightarrow \sigma) \Rightarrow \left(\overset{\pi}{\neg}\sigma \Rightarrow \overset{\pi}{\neg}\tau\right)$
$\overset{\pi}{\neg}\sigma \vee \overset{\pi}{\neg}\overset{\pi}{\neg}\sigma$	$\left(\sigma \Rightarrow \overset{\pi}{\neg}\sigma\right) \Leftrightarrow \overset{\pi}{\neg}\sigma$
$\left(\tau \Rightarrow \overset{\pi}{\neg}\sigma\right) \Rightarrow \left((\tau \Rightarrow \sigma) \Rightarrow \left(\overset{\pi}{\neg}\tau\right)\right)$	$\left(\tau \Rightarrow \overset{\pi}{\neg}\sigma\right) \Rightarrow \overset{\pi}{\neg}(\tau \wedge \sigma)$
$\left(\tau \Rightarrow \overset{\pi}{\neg}\sigma\right) \Leftrightarrow \left(\overset{\pi}{\neg}\tau \vee \overset{\pi}{\neg}\sigma\right)$	$\left(\tau \Rightarrow \overset{\pi}{\neg}\sigma\right) \Rightarrow \left((\tau \Rightarrow \sigma) \Rightarrow \left(\overset{\pi}{\neg}\tau\right)\right)$
$\overset{\pi}{\neg}\overset{\pi}{\neg}\left(\sigma \vee \overset{\pi}{\neg}\sigma\right)$	$\sigma \vee \tau \Rightarrow \left(\overset{\pi}{\neg}\sigma \Rightarrow \overset{\pi}{\neg}\overset{\pi}{\neg}\tau\right)$
$\neg\neg(\neg\neg\sigma \Rightarrow \sigma)$	$\left(\sigma \wedge \overset{\pi}{\neg}\tau\right) \Rightarrow \overset{\pi}{\neg}(\sigma \Rightarrow \tau)$
$\overset{\pi}{\neg}(\sigma \vee \tau) \Leftrightarrow \overset{\pi}{\neg}\sigma \wedge \overset{\pi}{\neg}\tau$	$(\neg\sigma \vee \tau) \Rightarrow (\sigma \Rightarrow \tau)$
$\overset{\pi}{\neg}\overset{\pi}{\neg}(\tau \wedge \sigma) \Rightarrow \left(\overset{\pi}{\neg}\overset{\pi}{\neg}\tau \wedge \overset{\pi}{\neg}\overset{\pi}{\neg}\sigma\right)$	$\left(\tau \Rightarrow \overset{\pi}{\neg}\sigma\right) \vee \left(\varphi \Rightarrow \overset{\pi}{\neg}\overset{\pi}{\neg}\sigma\right)$
$\left(\overset{\pi}{\neg}\sigma \vee \overset{\pi}{\neg}\tau\right) \Rightarrow \overset{\pi}{\neg}(\sigma \wedge \tau)$	$\overset{\pi}{\neg}\sigma \vee \left(\overset{\pi}{\neg}\sigma \Rightarrow \overset{\pi}{\neg}\tau\right)$
$\sigma \Rightarrow \left(\overset{\pi}{\neg}\sigma \Rightarrow \overset{\pi}{\neg}\tau\right)$	$\overset{\pi}{\neg}\overset{\pi}{\neg}(\sigma \vee \tau) \Leftrightarrow \left(\overset{\pi}{\neg}\overset{\pi}{\neg}\sigma \vee \overset{\pi}{\neg}\overset{\pi}{\neg}\tau\right)$
$((\sigma \Rightarrow 0) \vee \tau) \Rightarrow (\sigma \Rightarrow \tau)$	$(\sigma \vee \tau) \Rightarrow \overset{\pi}{\neg}\left(\overset{\pi}{\neg}\sigma \wedge \overset{\pi}{\neg}\tau\right)$
$\neg\sigma \Rightarrow (\sigma \Rightarrow \tau)$	$\partial_\pi\sigma \wedge \overset{\pi}{\neg}\overset{\pi}{\neg}\sigma \Leftrightarrow \sigma \vee \pi$
$\overset{\pi}{\neg}\overset{\pi}{\neg}(\sigma \Rightarrow \tau) \Rightarrow \left(\overset{\pi}{\neg}\overset{\pi}{\neg}\sigma \Rightarrow \overset{\pi}{\neg}\overset{\pi}{\neg}\tau\right)$	$\partial_\pi(\sigma \vee \tau) \Leftrightarrow (\partial_\pi\sigma \vee \tau) \wedge (\sigma \vee \partial_\pi\tau)$
$\left(\tau \Rightarrow \overset{\pi}{\neg}\sigma\right) \Leftrightarrow \left(\sigma \Rightarrow \overset{\pi}{\neg}\tau\right)$	$\left(\overset{\pi}{\neg}\sigma \Rightarrow \overset{\pi}{\neg}\tau\right) \vee \left(\overset{\pi}{\neg}\tau \Rightarrow \overset{\pi}{\neg}\sigma\right)$

Table 2.7: Some partition tautologies involving π-negation

Table 2.8 lists a few more partition tautologies involving π-negation.

$\overset{\pi}{\neg}0_U$	$\sigma \wedge \left(\sigma \vee \overset{\pi}{\neg}\sigma\right) \Leftrightarrow \sigma$
$\overset{\pi}{\neg}\sigma \vee \left(\overset{\pi}{\neg}\sigma \Uparrow \sigma\right) \Leftrightarrow \sigma$	$\sigma \wedge (\tau \vee (\sigma \Uparrow \pi)) \Leftrightarrow \sigma \wedge (\tau \vee \pi)$
$\left(\overset{\pi}{\neg}\sigma \vee (\sigma \vee \pi)\right) \Leftrightarrow \left(\overset{\pi}{\neg}\sigma \vee \sigma\right) = \partial_\pi \sigma$	$\tau \vee (\sigma \Uparrow \pi) \Uparrow (\sigma \Uparrow (\tau \vee \pi))$
$\left(\overset{\pi}{\neg}\sigma \vee \tau\right) \Leftrightarrow \left(\overset{\pi}{\neg}\sigma \Uparrow (\tau \vee \pi)\right)$	$(\sigma \Uparrow \pi) \vee (\sigma \Uparrow \tau) \Uparrow (\sigma \Uparrow (\tau \vee \pi))$
$\overset{\pi}{\neg}\sigma \vee \tau \Uparrow \overset{\pi}{\neg}\sigma$	$\overset{\pi}{\neg}(\pi \Uparrow (\pi \wedge \sigma \wedge \tau)) \Uparrow (\overset{\pi}{\neg}\pi \Uparrow (\pi \wedge \sigma)) \vee (\overset{\pi}{\neg}\pi \Uparrow (\pi \wedge \tau))$
$\overset{\pi}{\neg}\sigma \Uparrow \left(\overset{\pi}{\neg}(\sigma \vee \tau)\right)$	$\pi \vee (\pi \Uparrow (\pi \wedge \sigma)) \Leftrightarrow \overset{\pi}{\neg}\pi \Uparrow (\pi \wedge \sigma)$
$\tau \Uparrow [(\tau \Uparrow \sigma) \Uparrow ((\sigma \Uparrow \pi) \Uparrow \pi)]$	$\varphi \vee \left(\varphi \vee \overset{\pi}{\neg}\sigma\right) \Leftrightarrow \varphi$
$\tau \Uparrow [(\tau \Uparrow (\sigma \wedge \pi)) \Uparrow ((\tau \Uparrow \sigma) \Uparrow \pi)]$	$\overset{\pi}{\neg}(\tau \Uparrow \sigma)$
$(\sigma \Uparrow \tau) \Uparrow (\overset{\pi}{\neg}\tau \vee \sigma)$	$(\sigma \vee \tau) \Uparrow \overset{\pi}{\neg}\overset{\pi}{\neg}\tau$
$\overset{\pi}{\neg}\sigma \Uparrow (\overset{\pi}{\neg}\sigma \vee \tau)$	$(\sigma \vee \tau) \Uparrow \overset{\pi}{\neg}(\sigma \vee \tau \vee \pi)$
$\overset{\pi}{\neg}\left(\overset{\pi}{\neg}(\sigma \Uparrow \tau) \wedge \tau\right)$	$\overset{\pi}{\neg}\sigma \vee (\sigma \vee \pi)$
$\overset{\pi}{\neg}\sigma \Uparrow \overset{\pi}{\neg}\sigma$	$\tau \Leftrightarrow ((\tau \vee \sigma) \wedge (\tau \Uparrow \sigma)) \vee (\sigma \Uparrow \tau)$
$\overset{\pi}{\neg}\sigma \Uparrow \overset{\pi}{\neg}\sigma$	$\sigma \wedge \tau \Leftrightarrow (\sigma \vee \tau) \wedge (\tau \Uparrow \sigma) \vee (\sigma \Uparrow \tau)$
$\sigma \Uparrow \left(\tau \Uparrow \overset{\pi}{\neg}\overset{\pi}{\neg}\sigma \wedge (\tau \vee \pi)\right)$	$\overset{\pi}{\neg}\sigma \Uparrow (\sigma \vee \pi)$
	$\overset{\pi}{\neg}\overset{\pi}{\neg}\pi$

Table 2.8: More partition tautologies involving π-negation

Table 2.9 is a list of some partition tautologies that not intuitionistically valid and vice-versa.

Partition tautologies not intuit. tautologies	Intuit. tautologies not part. tautologies
$\overset{\pi}{\neg}\sigma \vee \overset{\pi}{\neg}\overset{\pi}{\neg}\sigma$	$\sigma \Rightarrow (\pi \Rightarrow (\sigma \wedge \pi))$
$\left(\tau \Rightarrow \overset{\pi}{\neg}\sigma\right) \vee \left(\varphi \Rightarrow \overset{\pi}{\neg}\overset{\pi}{\neg}\sigma\right)$	$\neg\neg[\sigma \Rightarrow (\pi \Rightarrow (\sigma \wedge \pi))]$
$\overset{\pi}{\neg}\overset{\pi}{\neg}(\sigma \vee \tau) \Rightarrow \left(\overset{\pi}{\neg}\overset{\pi}{\neg}\sigma \vee \overset{\pi}{\neg}\overset{\pi}{\neg}\tau\right)$	$((\pi \vee \sigma) \wedge \tau) \Rightarrow ((\pi \wedge \tau) \vee (\sigma \wedge \tau))$
$\left(\tau \Rightarrow \overset{\pi}{\neg}\sigma\right) \equiv \left(\overset{\pi}{\neg}\tau \vee \overset{\pi}{\neg}\sigma\right)$	$((\tau \Rightarrow \sigma) \wedge (\tau \Rightarrow \pi)) \Rightarrow (\tau \Rightarrow (\sigma \wedge \pi))$
$\left(\overset{\pi}{\neg}\sigma \Rightarrow \overset{\pi}{\neg}\tau\right) \vee \left(\overset{\pi}{\neg}\tau \Rightarrow \overset{\pi}{\neg}\sigma\right)$	$(\neg\neg\tau \wedge \neg\neg\sigma) \Rightarrow \neg\neg(\tau \wedge \sigma)$
$\left(\overset{\pi}{\neg}\sigma \vee \tau\right) = \left(\overset{\pi}{\neg}\overset{\pi}{\neg}\sigma \Rightarrow (\tau \vee \pi)\right)$	$((\tau \wedge \sigma) \Rightarrow \pi) \Rightarrow (\tau \Rightarrow (\sigma \Rightarrow \pi))$
$\left[\left(\tau \Rightarrow \overset{\pi}{\neg}\sigma\right) \Rightarrow \overset{\pi}{\neg}\tau\right] \vee \left[(\tau \Rightarrow \sigma) \Rightarrow \overset{\pi}{\neg}\tau\right]$	$(\sigma \Rightarrow \pi) \Rightarrow ((\sigma \wedge \tau) \Rightarrow (\pi \wedge \tau))$

Table 2.9: Some partition tautologies that are not intuitionistically valid and vice-versa

2.6　Towards an axiom system for partition logic

There is a semantic tableau system for partition logic [24] with a consistency proof and a (tedious) completeness argument. But there is not as yet a Hilbert-style axiom system that allows the derivation of all and only the partition tautologies. Our purpose here is only to give a suggested set of rules of inference and axioms, and to show how the derivations are carried out. The connectives used are the lattice operations of join $\vee$ and meet $\wedge$, the implication $\Rightarrow$, and the top 1 and bottom 0 (instead of the previous $\mathbf{1}_U$ and $\mathbf{0}_U$). Axioms and theorems are indicated by the turnstile $\vdash$ as in $\vdash \Phi$. The non-trivial axioms are introduced successively to derive more theorems. The bi-implication $\pi \Leftrightarrow \sigma$ is short for $(\pi \Rightarrow \sigma) \wedge (\sigma \Rightarrow \pi)$. The Boolean core L_π is the set of formulas of the form $\sigma \Rightarrow \pi$ for fixed π, the π-regular formulas.

　　Rule of inference 1: Modus ponens. If $\vdash \sigma$ and $\vdash \sigma \Rightarrow \pi$, then $\vdash \pi$.

–

Top Axioms:
Top of order: $\vdash \pi \Rightarrow 1$
Top and join: $\vdash \pi \vee 1$
Top and meet: $\vdash \pi \Rightarrow (\pi \wedge 1)$

–

Bottom Axioms:
Bottom of order: $\vdash 0 \Rightarrow \pi$
Bottom and join: $\vdash (\pi \vee 0) \Rightarrow \pi$
Bottom and meet: $\vdash 0 \Rightarrow (\pi \wedge 0)$

Join Axioms:
Commutativity: $\vdash (\pi \vee \sigma) \Leftrightarrow (\sigma \vee \pi)$.
Associativity: $\vdash (\tau \vee (\sigma \vee \pi)) \Leftrightarrow ((\tau \vee \sigma) \vee \pi)$
Upper bound: $\vdash \pi \Rightarrow (\sigma \vee \pi)$
Absorption: $\vdash \pi \vee \pi \Rightarrow \pi$.

Meet Axioms:
Commutativity: $\vdash (\pi \wedge \sigma) \Leftrightarrow (\sigma \wedge \pi)$
Associativity: $\vdash (\tau \wedge (\sigma \wedge \pi)) \Leftrightarrow ((\tau \wedge \sigma) \wedge \pi)$
Lower bound: $\vdash (\sigma \wedge \pi) \Rightarrow \sigma$
Absorption: $\vdash \pi \Rightarrow (\pi \wedge \pi)$.

Axiom 1 (Transitivity of Implication ToI)

$$\vdash (\tau \Rightarrow \sigma) \Rightarrow ((\sigma \Rightarrow \pi) \Rightarrow (\tau \Rightarrow \pi)).$$

Using the notation for π-negation $\overset{\pi}{\neg}\sigma = \sigma \Rightarrow \pi$, this axiom is just the contraposition for π-negation: $(\tau \Rightarrow \sigma) \Rightarrow \left(\overset{\pi}{\neg}\sigma \Rightarrow \overset{\pi}{\neg}\tau\right)$.

Theorem 18 $\vdash 1$.

Proof: $\vdash (1 \Rightarrow 1) \Rightarrow 1$ is an instance of the top axiom $\vdash \pi \Rightarrow 1$ with with π replaced by $1 \Rightarrow 1$, and then $\pi = 1$ again in $\vdash \pi \Rightarrow 1$ gives $\vdash 1 \Rightarrow 1$, so by modus ponens, the theorem follows. □

Theorem 19 $\vdash \pi \Rightarrow \pi$. *(reflexivity of implication)*

Proof: $\vdash \pi \Rightarrow (\pi \vee \pi)$ by join upper bound and $\vdash \pi \vee \pi \Rightarrow \pi$ by join absorption. Then substituting into the transitivity of implication, $\vdash (\pi \Rightarrow (\pi \vee \pi)) \Rightarrow (((\pi \vee \pi) \Rightarrow \pi) \Rightarrow (\pi \Rightarrow \pi))$, and then using modus ponens from the upper bound premise, $\vdash (((\pi \vee \pi) \Rightarrow \pi) \Rightarrow (\pi \Rightarrow \pi))$, and the modus ponens from the absorption premise, gives $\vdash \pi \Rightarrow \pi$. □

Once we get accustomed to the ToI axiom, then in a proof once we have proved $\sigma \Rightarrow \pi$ and $\tau \Rightarrow \sigma$, then we can just derive $\tau \Rightarrow \pi$ using it in effect as a derived rule of inference.

Theorem 20 $\vdash \pi \Rightarrow (\pi \vee \pi)$.

Proof: Upper bound with σ replaced by π. □

Theorem 21 $\vdash \pi \Rightarrow \pi \vee \sigma$.

Proof: $\vdash \sigma \vee \pi \Rightarrow \pi \vee \sigma$ by commutativity of join and $\pi \Rightarrow \sigma \vee \pi$ by upper bound, so by transitivity of implication, we have $\vdash \pi \Rightarrow (\pi \vee \sigma)$. □

These proofs should also be repeated for the meet since that is an independent connective in partition logic.

Theorem 22 $\vdash \pi \wedge \pi \Rightarrow \pi$.

Proof: By lower bound by substitution. □

Theorem 23 $\vdash \pi \wedge \sigma \Rightarrow \sigma$.

Proof: $\vdash \sigma \wedge \pi \Rightarrow \sigma$ by lower bound and $\vdash \pi \wedge \sigma \Rightarrow \sigma \wedge \pi$ by commutativity of meet so by the ToI, $\vdash \pi \wedge \sigma \Rightarrow \sigma$. □

Axiom 2 (Modus Ponens MP) $\vdash \sigma \wedge (\sigma \Rightarrow \pi) \Leftrightarrow (\sigma \wedge \pi)$.

Rule of inference 2: If $\vdash \sigma \Rightarrow \pi$, then $\vdash (\sigma \wedge \tau) \Rightarrow (\pi \wedge \tau)$. (monotony of meets)

Theorem 24 *If* $\vdash \sigma \Rightarrow \pi$, *then* $\vdash \sigma \Leftrightarrow \sigma \wedge \pi$. *(meet equivalent of ordering)*

Proof: $\vdash \sigma \wedge \sigma \Rightarrow (\sigma \wedge \pi)$ by monotony of meets and $\vdash \sigma \Rightarrow (\sigma \wedge \sigma)$ so $\vdash \sigma \Rightarrow (\sigma \wedge \pi)$ and the reverse implication follows by the lower bound property of the meet. □

Theorem 25 $\vdash \sigma \wedge (\sigma \Rightarrow \pi) \Rightarrow \pi$. *(internal modus ponens)*

Proof: We have $\vdash [\sigma \wedge (\sigma \Rightarrow \pi) \Leftrightarrow (\sigma \wedge \pi)] \Rightarrow (\sigma \wedge (\sigma \Rightarrow \pi) \Rightarrow (\sigma \wedge \pi))$ by meet lower bound so by modus ponens rule of inference and MP axiom, we have $\vdash \sigma \wedge (\sigma \Rightarrow \pi) \Rightarrow (\sigma \wedge \pi)$. Then using $\vdash \sigma \wedge \pi \Rightarrow \pi$ so by the ToI, we have $\vdash \sigma \wedge (\sigma \Rightarrow \pi) \Rightarrow \pi$. □

Theorem 26 $\vdash \sigma \Rightarrow (1 \Rightarrow \sigma)$.

Proof: Using the MP axiom, $\vdash 1 \wedge \sigma \Rightarrow 1 \wedge (1 \Rightarrow \sigma)$ and $\vdash \sigma \Rightarrow 1 \wedge \sigma$ by the top and meet axiom and $\vdash (1 \wedge (1 \Rightarrow \sigma)) \Rightarrow ((1 \Rightarrow \sigma))$ by meet lower bound so by ToI and some MP, we have the theorem. □

Theorem 27 $\vdash (1 \Rightarrow \sigma) \Rightarrow \sigma$.

Proof: $\vdash 1 \wedge (1 \Rightarrow \sigma) \Rightarrow 1 \wedge \sigma$ by MP axiom and $\vdash (1 \Rightarrow \sigma) \Rightarrow 1 \wedge (1 \Rightarrow \sigma)$ and $\vdash 1 \wedge \sigma \Rightarrow \sigma$ so by the ToI, the result follows. $\square$

Theorem 28 *If $\vdash \sigma$ and $\vdash \tau$, then $\vdash \sigma \wedge \tau$. (derived rule of inference)*

Proof: If σ is a theorem, then by the above theorem and modus ponens, $1 \Rightarrow \sigma$ is a theorem so by the monotony of meets rule of inference, $1 \wedge \tau \Rightarrow (\sigma \wedge \tau)$ is a theorem. By top and meet axiom, $\vdash \tau \Rightarrow (1 \wedge \tau)$ so using that τ is a theorem and modus ponens, we have $1 \wedge \tau$ and then modus ponens again yields $\sigma \wedge \tau$.$\square$

Theorem 29 $\vdash (1 \Rightarrow \sigma) \Leftrightarrow \sigma$.

Proof: By the previous three theorems. $\square$

Theorem 30 $\vdash (\sigma \wedge \tau) \Leftrightarrow (\sigma \wedge (\sigma \Rightarrow (\sigma \wedge \tau)))$.

Proof: Taking $\pi = \sigma \wedge \tau$, we have using lower bound and modus ponens axiom, $\vdash \sigma \wedge (\sigma \wedge \tau) \Rightarrow (\sigma \wedge (\sigma \Rightarrow (\sigma \wedge \tau)))$. By the associativity of the meet, $\vdash (\sigma \wedge \sigma) \wedge \tau \Rightarrow (\sigma \wedge (\sigma \wedge \tau))$ so by modus ponens, $\vdash (\sigma \wedge \sigma) \wedge \tau \Rightarrow (\sigma \wedge (\sigma \Rightarrow (\sigma \wedge \tau)))$. By absorption, $\vdash \sigma \Rightarrow (\sigma \wedge \sigma)$ and by monotony for meets rule of inference, $\vdash \sigma \wedge \tau \Rightarrow (\sigma \wedge \sigma) \wedge \tau$ so the result $\vdash (\sigma \wedge \tau) \Leftrightarrow (\sigma \wedge (\sigma \Rightarrow (\sigma \wedge \tau)))$ follows by modus ponens. Then by the modus ponens axiom again, $\vdash \sigma \wedge (\sigma \Rightarrow (\sigma \wedge \tau)) \Rightarrow \sigma \wedge (\sigma \wedge \tau)$ so by associativity, $\vdash (\sigma \wedge (\sigma \wedge \tau)) \Rightarrow (\sigma \wedge \sigma) \wedge \tau$. Then by upper bound $\vdash \sigma \wedge \sigma \Rightarrow \sigma$ and the meet monotony rule of inference, $\vdash (\sigma \wedge \sigma) \wedge \tau \Rightarrow \sigma \wedge \tau$ so with two applications of modus ponens, the implication holds in the other direction. But to show the biconditional, we need some method to prove a conjunction or meet, and that is supplied by the previous theorem giving the derived rule of inference to do the job. $\square$

Theorem 31 $\vdash \sigma \wedge (\sigma \Rightarrow 1) \Leftrightarrow \sigma$.

Proof: By the MP axiom, $\vdash \sigma \wedge (\sigma \Rightarrow 1) \Leftrightarrow \sigma \wedge 1$ and $\vdash \sigma \wedge 1 \Rightarrow \sigma$ by the meet lower bound and $\vdash \sigma \Rightarrow \sigma \wedge 1$ by top and meet axiom, so using the ToI, the result follows, i.e., the implication each way is proven and then the theorem follows by the derived rule of inference. $\square$

Theorem 32 $\vdash \sigma \wedge (\sigma \Rightarrow (\pi \wedge \tau)) \Leftrightarrow \sigma \wedge (\sigma \Rightarrow \pi) \wedge (\sigma \Rightarrow \tau)$. *(RAPL)*

Proof: By the MP axiom, $\vdash \sigma \wedge (\sigma \Rightarrow (\pi \wedge \tau)) \Leftrightarrow \sigma \wedge \pi \wedge \tau$. But $\vdash \sigma \wedge (\sigma \Rightarrow \pi) \Leftrightarrow \sigma \wedge \pi$ and then by the monotony of meets rule of inference, $\vdash \sigma \wedge \pi \wedge \tau \Rightarrow \sigma \wedge (\sigma \Rightarrow \pi) \wedge \tau$ and vice versa so we have the implication each way between the LHS and $\vdash \sigma \wedge (\sigma \Rightarrow \pi) \wedge \tau$. Then using commutativity of meet, associativity of meet, and the MP axiom with the monotony of meets, we have $\vdash \sigma \wedge (\sigma \Rightarrow \pi) \wedge \tau$ implies $\vdash \sigma \wedge (\sigma \Rightarrow \pi) \wedge (\sigma \Rightarrow \tau)$ and vice versa. Thus we have the two implications of the theorem so the bi-implication follows by the derived rule of inference.   □

Category theorists might interpret that theorem as "Right Adjoints Preserve Limits" or RAPL.

Theorem 33 *If $\vdash \sigma \Rightarrow \pi$ and $\vdash \sigma \Leftrightarrow \tau$, and $\vdash \pi \Leftrightarrow \phi$, then $\vdash \tau \Rightarrow \phi$.*

Proof: Clearly, $\vdash \sigma \Rightarrow \phi$ is a theorem by the transitivity of implication and modus ponens, and then $\tau \Rightarrow \phi$ follows by the same argument.   □

Theorem 34 *If $\vdash \tau \Rightarrow \pi$, then $\vdash \sigma \wedge (\sigma \Rightarrow \tau) \Rightarrow (\sigma \wedge (\sigma \Rightarrow \pi))$.*

Proof: Since $\vdash \tau \Rightarrow \pi$ is a theorem, by monotony of meets, $\vdash \tau \wedge \sigma \Rightarrow \pi \wedge \sigma$ is a theorem and each can be replaced by the bi-implication equivalent to yield the theorem.   □

Theorem 35 *$\vdash \sigma \Rightarrow \pi$ implies $\vdash \sigma \Leftrightarrow (\sigma \wedge (\sigma \Rightarrow \pi))$.*

Proof: The RHS implies the LHS in general so we have to show that $\vdash \sigma \Rightarrow (\sigma \wedge (\sigma \Rightarrow \pi))$. We have from the modus ponens axiom that $\vdash \sigma \wedge \pi \Rightarrow \sigma \wedge (\sigma \Rightarrow \pi)$. The assumption $\vdash \sigma \Rightarrow \pi$ yields by the monotony of meet rule of inference that $\vdash \sigma \wedge \sigma \Rightarrow \pi \wedge \sigma$ and we already have $\vdash \sigma \Rightarrow \sigma \wedge \sigma$ so by the ToI, we have the other implication, and then the result follows by the previous derived rule of inference.   □

Theorem 36 *If $\vdash \pi \Rightarrow \sigma$ then $\sigma \wedge \overset{\pi}{\neg}\sigma \Leftrightarrow \pi$.*

Proof: By the MP axiom, $\vdash \sigma \wedge \overset{\pi}{\neg}\sigma \Leftrightarrow \sigma \wedge \pi$. Now $\vdash \sigma \wedge \pi \Rightarrow \pi$ so we have the left to right implication by modus ponens independent of the assumption. Using the assumption and monotony of the meet and absorption, we have $\vdash \pi \Rightarrow \sigma \wedge \pi$ which gives the implication the other way.   □

The notation $\neg \sigma$ is an abbreviation for $\sigma \Rightarrow 0$.

Theorem 37 *$\sigma \wedge \neg \sigma \Leftrightarrow 0$.*

Proof: Since $0 \Rightarrow \sigma$ is an axiom, the theorem follows by the previous result.   □

Lemma 38 *If $\vdash \varphi$ then $\vdash \tau \Leftrightarrow \varphi \wedge \tau$. (any theorem can replace 1 in $\tau \Rightarrow 1 \wedge \tau$)*

Proof: $\vdash \varphi \Rightarrow (1 \Rightarrow \varphi)$ and $\vdash \varphi$ so $\vdash 1 \Rightarrow \varphi$. Then applying monotony of meets, $\vdash 1 \wedge \tau \Rightarrow \varphi \wedge \tau$ and $\vdash \tau \Rightarrow (1 \wedge \tau)$ so by transitivity of implication, $\vdash \tau \Rightarrow \varphi \wedge \tau$. The reverse holds in general by lower bound of meet axiom. $\qquad\square$

Lemma 39 *If $\vdash \varphi$ then $\vdash (\varphi \Rightarrow \sigma) \Rightarrow \sigma$.*

Proof: By the MP axiom, $\vdash \varphi \wedge (\varphi \Rightarrow \sigma) \Rightarrow (\varphi \wedge \sigma)$, and by the previous lemma $\vdash (\varphi \Rightarrow \sigma) \Rightarrow \varphi \wedge (\varphi \Rightarrow \sigma)$ so by the transitivity of implication, $\vdash (\varphi \Rightarrow \sigma) \Rightarrow (\varphi \wedge \sigma)$. Then the result follows since $\vdash (\varphi \wedge \sigma) \Rightarrow \sigma$ by lower bound. $\qquad\square$

Theorem 40 $\vdash \sigma \Rightarrow \overset{\pi}{\neg}\overset{\pi}{\neg}\sigma$. *(one-way double π-negation)*

Proof: Plugging into transitivity of implication, we have:

$$\vdash (1 \Rightarrow \sigma) \Rightarrow ((\sigma \Rightarrow \pi) \Rightarrow (1 \Rightarrow \pi))$$

and to simplify the consequent, we again plug into ToI:

$$\vdash ((\sigma \Rightarrow \pi) \Rightarrow (1 \Rightarrow \pi)) \Rightarrow (((1 \Rightarrow \pi) \Rightarrow \pi) \Rightarrow ((\sigma \Rightarrow \pi) \Rightarrow \pi)).$$

We can then apply transitivity of implication and two modus ponens to get:

$$\vdash (1 \Rightarrow \sigma) \Rightarrow (((1 \Rightarrow \pi) \Rightarrow \pi) \Rightarrow ((\sigma \Rightarrow \pi) \Rightarrow \pi))$$

and since $\sigma \Rightarrow (1 \Rightarrow \sigma)$ is a theorem, we have:

$$\vdash \sigma \Rightarrow (((1 \Rightarrow \pi) \Rightarrow \pi) \Rightarrow ((\sigma \Rightarrow \pi) \Rightarrow \pi)).$$

Since $((1 \Rightarrow \pi) \Rightarrow \pi)$ is a theorem, we have by the lemma,

$$\vdash [((1 \Rightarrow \pi) \Rightarrow \pi) \Rightarrow ((\sigma \Rightarrow \pi) \Rightarrow \pi)] \Rightarrow ((\sigma \Rightarrow \pi) \Rightarrow \pi).$$

Then the result follows by the ToI. $\qquad\square$

Lemma 41 *If $\vdash \phi \Rightarrow (\varphi \Rightarrow \tau)$ and $\vdash \psi \Rightarrow \varphi$, then $\vdash \phi \Rightarrow (\psi \Rightarrow \tau)$.*

Proof: From $\vdash \psi \Rightarrow \varphi$ and ToI, we have $\vdash (\varphi \Rightarrow \tau) \Rightarrow (\psi \Rightarrow \tau)$. Then the result follows from ToI and a couple of modus ponens rules. $\qquad\square$

Theorem 42 $\vdash (\tau \Rightarrow (\sigma \Rightarrow \pi)) \Leftrightarrow (\sigma \Rightarrow (\tau \Rightarrow \pi))$. *(interchange)*

Proof: Plugging into the ToI gives:

$$\vdash (\tau \Rightarrow (\sigma \Rightarrow \pi)) \Rightarrow [((\sigma \Rightarrow \pi) \Rightarrow \pi) \Rightarrow (\tau \Rightarrow \pi)].$$

By the double negation theorem, $\vdash \sigma \Rightarrow ((\sigma \Rightarrow \pi) \Rightarrow \pi)$, so by the previous lemma, $\vdash (\tau \Rightarrow (\sigma \Rightarrow \pi)) \Rightarrow (\sigma \Rightarrow (\tau \Rightarrow \pi))$. By symmetry, the reverse implication holds to the bi-implication follows. $\square$

Corollary 15 $\vdash \left(\overset{\pi}{\neg}\tau \Rightarrow \overset{\pi}{\neg}\sigma \right) \Leftrightarrow \left(\sigma \Rightarrow \overset{\pi}{\neg}\overset{\pi}{\neg}\tau \right)$.

Proof: Plug $\overset{\pi}{\neg}\tau$ in for τ in the interchange theorem. $\square$

Theorem 43 $\vdash \neg\sigma \Rightarrow (\sigma \Rightarrow \tau)$.

Proof: The ToI gives: $\vdash (\sigma \Rightarrow 0) \Rightarrow ((0 \Rightarrow \tau) \Rightarrow (\sigma \Rightarrow \tau))$ and then the interchange gives $\vdash (0 \Rightarrow \tau) \Rightarrow ((\sigma \Rightarrow 0) \Rightarrow (\sigma \Rightarrow \tau))$ where the premise is an axiom so the result follows. $\square$

Theorem 44 $\overset{\pi}{\neg}\overset{\pi}{\neg}\sigma \wedge \left(\overset{\pi}{\neg}\overset{\pi}{\neg}\sigma \Rightarrow \sigma \right) \Leftrightarrow \sigma$.

Proof: By the MP axiom, $\vdash \overset{\pi}{\neg}\overset{\pi}{\neg}\sigma \wedge \left(\overset{\pi}{\neg}\overset{\pi}{\neg}\sigma \Rightarrow \sigma \right) \Leftrightarrow \sigma \wedge \overset{\pi}{\neg}\overset{\pi}{\neg}\sigma$. Since $\vdash \sigma \Rightarrow \overset{\pi}{\neg}\overset{\pi}{\neg}\sigma$ we have by monotony of meets, $\sigma \wedge \sigma \Rightarrow \overset{\pi}{\neg}\overset{\pi}{\neg}\sigma \wedge \sigma$ and then by absorption, $\sigma \Rightarrow \overset{\pi}{\neg}\overset{\pi}{\neg}\sigma \wedge \sigma$. The reverse implication is an instance of lower bound, so we have the implication each way and then the theorem follows by ToI. $\square$

Lemma 45 *If $\vdash \varphi \Rightarrow \tau$ and $\vdash \varphi \Rightarrow \sigma$, then $\vdash \varphi \Rightarrow \tau \wedge \sigma$. (lub property of meet)*

Proof: By monotony of meets, $\vdash \varphi \wedge \sigma \Rightarrow \tau \wedge \sigma$ and $\vdash \varphi \wedge \varphi \Rightarrow \varphi \wedge \sigma$ as well as $\vdash \varphi \Rightarrow \varphi \wedge \varphi$ so by the transitivity of implication, $\vdash \varphi \Rightarrow \tau \wedge \sigma$. $\square$

This is an important result because it suggests the successful internalization of the greatest lower bound property of the meet. It only uses monotony of meets and ToI. But the formula $\tau \Rightarrow (\sigma \Rightarrow \sigma \wedge \tau)$ is a subset tautology that is not valid in partition logic as one can check with $\tau = \{\{a\}, \{b, c\}\}$ and $\sigma = \{\{a, b\}, \{c\}\}$ so $\sigma \wedge \tau = \mathbf{0}_U$ and $\tau \Rightarrow (\sigma \Rightarrow \sigma \wedge \tau) = \mathbf{0}_U \neq \mathbf{1}_U$.

Lemma 46 $\vdash (\tau \Rightarrow \sigma) \Rightarrow \left(\overset{\pi}{\neg}\overset{\pi}{\neg}\tau \Rightarrow \overset{\pi}{\neg}\overset{\pi}{\neg}\sigma \right)$. *(covariance of double π-negation transform)*

Proof: By the ToI, $\vdash (\tau \Rightarrow \sigma) \Rightarrow ((\sigma \Rightarrow \pi) \Rightarrow (\tau \Rightarrow \pi))$ and again by ToI again, $\vdash ((\sigma \Rightarrow \pi) \Rightarrow (\tau \Rightarrow \pi)) \Rightarrow \left(\overset{\pi}{\neg}\overset{\pi}{\neg}\tau \Rightarrow \overset{\pi}{\neg}\overset{\pi}{\neg}\sigma \right)$ so the lemma follows by ToI and modus ponens. □

Theorem 47 $\vdash \overset{\pi}{\neg}\overset{\pi}{\neg}(\tau \wedge \sigma) \Rightarrow \left(\overset{\pi}{\neg}\overset{\pi}{\neg}\tau \wedge \overset{\pi}{\neg}\overset{\pi}{\neg}\sigma \right)$. *(reverse not a tautology)*

Proof: Since by lower bound, $\vdash \tau \wedge \sigma \Rightarrow \tau$ and $\vdash \tau \wedge \sigma \Rightarrow \sigma$ we can apply the previous double π-negation lemma and then the glb lemma to give the result.□

Theorem 48 $\vdash \overset{\pi}{\neg}\overset{\pi}{\neg}\left(\overset{\pi}{\neg}\sigma \wedge \sigma \right) \Leftrightarrow \pi$.

Proof: By the previous theorem, $\vdash \overset{\pi}{\neg}\overset{\pi}{\neg}\left(\overset{\pi}{\neg}\sigma \wedge \sigma \right) \Rightarrow \overset{\pi}{\neg}\overset{\pi}{\neg}\overset{\pi}{\neg}\sigma \wedge \overset{\pi}{\neg}\overset{\pi}{\neg}\sigma \Leftrightarrow \pi$ and the right to left implication is trivial. □

Theorem 49 $\vdash \overset{\pi}{\neg}\left(\overset{\pi}{\neg}\sigma \wedge \sigma \right)$. *(law of non-contradiction)*

Proof: Contrapositing the previous theorem gives:

$$\vdash (\pi \Rightarrow \pi) \Rightarrow \overset{\pi}{\neg}\overset{\pi}{\neg}\overset{\pi}{\neg}\left(\overset{\pi}{\neg}\sigma \wedge \sigma \right) \Rightarrow \overset{\pi}{\neg}\left(\overset{\pi}{\neg}\sigma \wedge \sigma \right)$$

and then modus ponens yields the result. Also, this is just a rewrite of the internal modus ponens. □

Theorem 50 $\vdash (\tau \wedge \sigma) \Rightarrow \pi$ *iff* $\vdash (\tau \wedge \sigma) \Rightarrow (\sigma \wedge (\sigma \Rightarrow \pi))$. *(internalized coreflection)*

Proof: Since $\vdash \tau \wedge \sigma \Rightarrow \sigma$, we have $\vdash \tau \wedge \sigma \Rightarrow \sigma \wedge \pi$ and $\vdash \sigma \wedge \pi \Rightarrow \sigma \wedge (\sigma \Rightarrow \pi)$ by the MP axiom so the RHS follows. Assuming the RHS, we have $\vdash \tau \wedge \sigma \Rightarrow \sigma \wedge \pi$ and then from $\vdash \sigma \wedge \pi \Rightarrow \pi$, the LHS follows. □

Theorem 51 $\vdash \pi \Rightarrow (\sigma \Rightarrow \pi)$.

Proof: Using ToI, $\vdash (\sigma \Rightarrow 1) \Rightarrow ((1 \Rightarrow \pi) \Rightarrow (\sigma \Rightarrow \pi))$ and $\vdash \sigma \Rightarrow 1$ is an axiom so the consequent follows. Also $\vdash \pi \Rightarrow (1 \Rightarrow \pi)$ so the result follows by the ToI. □

Theorem 52 $\vdash ((\sigma \Rightarrow \pi) \Rightarrow \sigma) \Rightarrow (\pi \Rightarrow \sigma)$. *(weakening of Pierce's law)*

Proof: Using the previous theorem and ToI,

$$\vdash (\pi \Rightarrow (\sigma \Rightarrow \pi)) \Rightarrow [((\sigma \Rightarrow \pi) \Rightarrow \sigma) \Rightarrow (\pi \Rightarrow \sigma)]$$

the result follows by modus ponens.   □

Theorem 53 $\vdash \overset{\pi}{\neg}\sigma \wedge \overset{\pi}{\neg}\overset{\pi}{\neg}\sigma \Leftrightarrow \pi$. *(law of contradiction in the Boolean core L_π)*

Proof: In the previous theorem, $\vdash \pi \Rightarrow \sigma$ then $\vdash \sigma \wedge \overset{\pi}{\neg}\sigma \Leftrightarrow \pi$, replace σ by $\overset{\pi}{\neg}\sigma$ and use the theorem that $\vdash \pi \Rightarrow \overset{\pi}{\neg}\sigma$.   □

Theorem 54 $\vdash \overset{\pi}{\neg}\overset{\pi}{\neg}\overset{\pi}{\neg}\sigma \Leftrightarrow \overset{\pi}{\neg}\sigma$. *(triple = single π-negation)*

Proof: $\vdash \overset{\pi}{\neg}\sigma \Rightarrow \overset{\pi}{\neg}\overset{\pi}{\neg}\overset{\pi}{\neg}\sigma$ by substitution in the double π-negation theorem. And then by contrapositing that theorem (i.e., ToI), we have the reverse implication. □

Theorem 55 $\vdash \sigma \Rightarrow \overset{\pi}{\neg}\tau$ iff $\vdash \overset{\pi}{\neg}\overset{\pi}{\neg}\sigma \Rightarrow \overset{\pi}{\neg}\tau$. *(double π-negation adjunction)*

Proof: Assuming the LHS, we have using the interchange and modus ponens, $\vdash \tau \Rightarrow \overset{\pi}{\neg}\sigma$ and then by contrapositing, we have $\vdash \overset{\pi}{\neg}\overset{\pi}{\neg}\sigma \Rightarrow \overset{\pi}{\neg}\tau$. Conversely we use $\vdash \sigma \Rightarrow \overset{\pi}{\neg}\overset{\pi}{\neg}\sigma$ and ToI to derive the LHS.   □

Theorem 56 $\vdash \left(\sigma \Rightarrow \overset{\pi}{\neg}\overset{\pi}{\neg}\tau\right) \Leftrightarrow \left(\overset{\pi}{\neg}\tau \Rightarrow \overset{\pi}{\neg}\sigma\right)$.

Proof: Apply interchange to the RHS.   □

Theorem 57 $\vdash \overset{\pi}{\neg}\overset{\pi}{\neg}\left(\overset{\pi}{\neg}\sigma \wedge \overset{\pi}{\neg}\tau\right) \Leftrightarrow \left(\overset{\pi}{\neg}\sigma \wedge \overset{\pi}{\neg}\tau\right)$. *(meet of π-regular elements is π-regular)*

Proof: $\vdash \overset{\pi}{\neg}\sigma \wedge \overset{\pi}{\neg}\tau \Rightarrow \overset{\pi}{\neg}\sigma, \overset{\pi}{\neg}\tau$ so applying double π-negation theorem, $\vdash \overset{\pi}{\neg}\overset{\pi}{\neg}\left(\overset{\pi}{\neg}\sigma \wedge \overset{\pi}{\neg}\tau\right) \overset{\pi}{\neg}\overset{\pi}{\neg}\overset{\pi}{\neg}\sigma, \overset{\pi}{\neg}\overset{\pi}{\neg}\overset{\pi}{\neg}\tau$ and then the triple negation theorem gives $\vdash \overset{\pi}{\neg}\overset{\pi}{\neg}\left(\overset{\pi}{\neg}\sigma \wedge \overset{\pi}{\neg}\tau\right) \Rightarrow \overset{\pi}{\neg}\sigma, \overset{\pi}{\neg}\tau$ and then the left to right implication follows. The right to left implication is a special case of the double π-negation implication so the theorem follows.   □

Theorem 58 $\vdash \sigma \Rightarrow \left(\overset{\pi}{\neg}\sigma \Rightarrow \overset{\pi}{\neg}\tau\right)$.

Proof: $\vdash (\sigma \Rightarrow \pi) \Rightarrow (\tau \Rightarrow (\sigma \Rightarrow \pi))$ by substituting in $\vdash \pi \Rightarrow (\sigma \Rightarrow \pi)$ and then using the interchange, we have: $\vdash (\sigma \Rightarrow \pi) \Rightarrow (\sigma \Rightarrow (\tau \Rightarrow \pi))$ and then another interchange applied to the whole formula gives the result.   □

Theorem 59 $\vdash \left(\overset{\pi}{\neg}\tau \Rightarrow \overset{\pi}{\neg}\sigma\right) \Leftrightarrow \left(\overset{\pi}{\neg}\overset{\pi}{\neg}\sigma \Rightarrow \overset{\pi}{\neg}\overset{\pi}{\neg}\tau\right)$. *(contraposition in the Boolean core L_π)*

Proof: The left to right implication follows by ordinary contraposition. Applying contraposition to the double π-negated formulas gives $\vdash \left(\overset{\pi}{\neg}\overset{\pi}{\neg}\sigma \Rightarrow \overset{\pi}{\neg}\overset{\pi}{\neg}\tau \right) \Rightarrow \left(\overset{\pi}{\neg}\overset{\pi}{\neg}\overset{\pi}{\neg}\tau \Rightarrow \overset{\pi}{\neg}\overset{\pi}{\neg}\overset{\pi}{\neg}\sigma \right)$.

Now $\vdash \left(\overset{\pi}{\neg}\tau \Rightarrow \overset{\pi}{\neg}\overset{\pi}{\neg}\overset{\pi}{\neg}\tau \right) \Rightarrow \left[\left(\overset{\pi}{\neg}\overset{\pi}{\neg}\overset{\pi}{\neg}\tau \Rightarrow \overset{\pi}{\neg}\overset{\pi}{\neg}\overset{\pi}{\neg}\sigma \right) \Rightarrow \left(\overset{\pi}{\neg}\tau \Rightarrow \overset{\pi}{\neg}\overset{\pi}{\neg}\overset{\pi}{\neg}\sigma \right) \right]$ and the premise is a theorem so we have $\vdash \left(\overset{\pi}{\neg}\overset{\pi}{\neg}\overset{\pi}{\neg}\tau \Rightarrow \overset{\pi}{\neg}\overset{\pi}{\neg}\overset{\pi}{\neg}\sigma \right) \Rightarrow \left(\overset{\pi}{\neg}\tau \Rightarrow \overset{\pi}{\neg}\overset{\pi}{\neg}\overset{\pi}{\neg}\sigma \right)$ and then by ToI, $\vdash \left(\overset{\pi}{\neg}\overset{\pi}{\neg}\sigma \Rightarrow \overset{\pi}{\neg}\overset{\pi}{\neg}\tau \right) \Rightarrow \left(\overset{\pi}{\neg}\tau \Rightarrow \overset{\pi}{\neg}\overset{\pi}{\neg}\overset{\pi}{\neg}\sigma \right)$. We also have

$$\vdash \left(\overset{\pi}{\neg}\tau \Rightarrow \overset{\pi}{\neg}\overset{\pi}{\neg}\overset{\pi}{\neg}\sigma \right) \Rightarrow \left(\left(\overset{\pi}{\neg}\overset{\pi}{\neg}\overset{\pi}{\neg}\sigma \Rightarrow \overset{\pi}{\neg}\sigma \right) \Rightarrow \left(\overset{\pi}{\neg}\tau \Rightarrow \overset{\pi}{\neg}\sigma \right) \right)$$

so by transitivity, $\vdash \left(\overset{\pi}{\neg}\overset{\pi}{\neg}\sigma \Rightarrow \overset{\pi}{\neg}\overset{\pi}{\neg}\tau \right) \Rightarrow \left(\left(\overset{\pi}{\neg}\overset{\pi}{\neg}\overset{\pi}{\neg}\sigma \Rightarrow \overset{\pi}{\neg}\sigma \right) \Rightarrow \left(\overset{\pi}{\neg}\tau \Rightarrow \overset{\pi}{\neg}\sigma \right) \right)$ so by interchange and using the triple π-negation theorem, the right to left implication follows. □

Corollary 16 *If* $\vdash \varphi$ *and* $\vdash \sigma \Rightarrow (\varphi \Rightarrow \pi)$, *then* $\vdash \sigma \Rightarrow \pi$.

Proof: Apply interchange and modus ponens. □

Corollary 17 $\vdash \left(\tau \Rightarrow \overset{\pi}{\neg}\sigma \right) \Leftrightarrow \left(\overset{\pi}{\neg}\overset{\pi}{\neg}\tau \Rightarrow \overset{\pi}{\neg}\sigma \right)$.

Proof: $\vdash \left(\tau \Rightarrow \overset{\pi}{\neg}\sigma \right) \Leftrightarrow \left(\sigma \Rightarrow \overset{\pi}{\neg}\tau \right) \Rightarrow \left(\overset{\pi}{\neg}\overset{\pi}{\neg}\tau \Rightarrow \overset{\pi}{\neg}\sigma \right)$.

Conversely, $\vdash \left(\overset{\pi}{\neg}\overset{\pi}{\neg}\tau \Rightarrow \overset{\pi}{\neg}\sigma \right) \Rightarrow \left(\overset{\pi}{\neg}\overset{\pi}{\neg}\sigma \Rightarrow \overset{\pi}{\neg}\overset{\pi}{\neg}\overset{\pi}{\neg}\tau \right)$ and $\vdash \left(\overset{\pi}{\neg}\overset{\pi}{\neg}\sigma \Rightarrow \overset{\pi}{\neg}\overset{\pi}{\neg}\overset{\pi}{\neg}\tau \right) \Rightarrow \left(\overset{\pi}{\neg}\overset{\pi}{\neg}\sigma \Rightarrow \overset{\pi}{\neg}\tau \right)$ so $\vdash \left(\overset{\pi}{\neg}\overset{\pi}{\neg}\tau \Rightarrow \overset{\pi}{\neg}\sigma \right) \Rightarrow \left(\overset{\pi}{\neg}\overset{\pi}{\neg}\sigma \Rightarrow \overset{\pi}{\neg}\tau \right)$ and $\vdash \left(\overset{\pi}{\neg}\overset{\pi}{\neg}\sigma \Rightarrow \overset{\pi}{\neg}\tau \right) \Rightarrow \left(\tau \Rightarrow \overset{\pi}{\neg}\overset{\pi}{\neg}\overset{\pi}{\neg}\sigma \right)$ by interchange and then $\vdash \left(\tau \Rightarrow \overset{\pi}{\neg}\overset{\pi}{\neg}\overset{\pi}{\neg}\sigma \right) \Rightarrow \left(\tau \Rightarrow \overset{\pi}{\neg}\sigma \right)$ by ToI and using triple negation theorem. The right to left implication follows by several uses of ToI. □

Axiom 3 (Monotony of join) $\vdash (\sigma \Rightarrow \pi) \Rightarrow ((\sigma \vee \tau) \Rightarrow (\pi \vee \tau))$.

Theorem 60 *If* $\vdash \sigma \Rightarrow \tau$, *then* $\vdash \tau \Leftrightarrow (\sigma \vee \tau)$. *(join equivalent of ordering)*

Proof: $\vdash \sigma \vee \tau \Rightarrow \tau \vee \tau \Leftrightarrow \tau$ and the reverse implication follows from the upper bound property of the join. □

Theorem 61 $\vdash \sigma \Rightarrow \pi$ *and* $\vdash \tau \Rightarrow \pi$ *iff* $\vdash (\sigma \vee \tau) \Rightarrow \pi$. *(glb property of join)*

Proof: Assuming the LHS, $\vdash \sigma \Rightarrow \pi$ implies $\vdash \sigma \vee \tau \Rightarrow \pi \vee \tau$ using monotony of join. And $\vdash \tau \Rightarrow \pi$ implies $\vdash \tau \vee \pi \Rightarrow \pi \vee \pi$ and thus $\vdash \tau \vee \pi \Rightarrow \pi$ so the RHS follows using ToI and modus ponens. Assuming the RHS, we use $\vdash \sigma \Rightarrow \sigma \vee \tau$ and $\vdash \tau \Rightarrow \sigma \vee \tau$ and ToI and MP to derive the LHS. $\square$

Theorem 62 $\vdash (\neg \sigma \vee \tau) \Rightarrow (\sigma \Rightarrow \tau)$, *i.e.*, $((\sigma \Rightarrow 0) \vee \tau) \Rightarrow (\sigma \Rightarrow \tau)$.

Proof: From $\vdash 0 \Rightarrow \tau$ and ToI, we have $\vdash (\sigma \Rightarrow 0) \Rightarrow (\sigma \Rightarrow \tau)$, which together with $\vdash \tau \Rightarrow (\sigma \Rightarrow \tau)$ gives by the previous theorem, $\vdash (\neg \sigma \vee \tau) \Rightarrow (\sigma \Rightarrow \tau)$. $\square$

Lemma 63 $\vdash \tau \vee (\sigma \Rightarrow \pi) \Rightarrow (\sigma \Rightarrow (\tau \vee \pi))$.

Proof: $\vdash \tau \Rightarrow (\tau \vee \pi)$ and $\vdash (\tau \vee \pi) \Rightarrow (\sigma \Rightarrow (\tau \vee \pi))$ so that $\vdash \tau \Rightarrow (\sigma \Rightarrow (\tau \vee \pi))$. Then from $\vdash \pi \Rightarrow (\tau \vee \pi)$ and ToI, we have $\vdash (\sigma \Rightarrow \pi) \Rightarrow (\sigma \Rightarrow (\tau \vee \pi))$, to by the previous derived rule of inference, the result follows. $\square$

Theorem 64 $\vdash \sigma \wedge (\tau \vee (\sigma \Rightarrow \pi)) \Leftrightarrow \sigma \wedge (\tau \vee \pi)$. *($\tau$-join relativized modus ponens)*

Proof: $\vdash \pi \Rightarrow (\sigma \Rightarrow \pi)$ implies $\vdash \tau \vee \pi \Rightarrow (\tau \vee (\sigma \Rightarrow \pi))$ by MoJ and MP, and then the right to left implication follows by MoM. By the MP axiom, $\vdash \sigma \wedge (\sigma \Rightarrow (\tau \vee \pi)) \Leftrightarrow \sigma \wedge (\tau \vee \pi)$. By the lemma, $\vdash \tau \vee (\sigma \Rightarrow \pi) \Rightarrow (\sigma \Rightarrow (\tau \vee \pi))$, and then by MoM $\vdash \sigma \wedge (\tau \vee (\sigma \Rightarrow \pi)) \Rightarrow (\sigma \wedge (\sigma \Rightarrow (\tau \vee \pi)))$ so by ToI, the left to right implication follows. $\square$

Theorem 65 $\vdash (\sigma \Rightarrow \pi) \vee (\sigma \Rightarrow \tau) \Rightarrow (\sigma \Rightarrow (\pi \vee \tau))$. *(internalized lub property of join)*

Proof: Since $\vdash \pi \Rightarrow (\pi \vee \tau)$ and $\vdash \tau \Rightarrow (\pi \vee \tau)$ so that $\vdash (\sigma \Rightarrow \pi) \Rightarrow (\sigma \Rightarrow \pi \vee \tau)$ and $\vdash (\sigma \Rightarrow \tau) \Rightarrow (\sigma \Rightarrow (\pi \vee \tau))$ and thus by the previous derived rule of inference, the result follows. $\square$

Theorem 66 (Weak DeMorgan) $\vdash \overset{\pi}{\neg}(\sigma \vee \tau) \Leftrightarrow \left(\overset{\pi}{\neg}\sigma \wedge \overset{\pi}{\neg}\tau\right)$, *i.e.*, $(\sigma \vee \tau) \Rightarrow \pi \Leftrightarrow (\sigma \Rightarrow \pi) \wedge (\tau \Rightarrow \pi)$.

Proof: $\vdash \sigma \Rightarrow \sigma \vee \tau$ and $\vdash \tau \Rightarrow \sigma \vee \tau$ so by contraposition, $\vdash \overset{\pi}{\neg}(\sigma \vee \tau) \Rightarrow \overset{\pi}{\neg}\sigma$ and $\vdash \overset{\pi}{\neg}(\sigma \vee \tau) \Rightarrow \overset{\pi}{\neg}\tau$ so by a previous theorem, $\vdash \overset{\pi}{\neg}(\sigma \vee \tau) \Rightarrow \left(\overset{\pi}{\neg}\sigma \wedge \overset{\pi}{\neg}\tau\right)$. Conversely, $\vdash \sigma \Rightarrow \overset{\pi}{\neg}\overset{\pi}{\neg}\sigma$ and $\vdash \tau \Rightarrow \overset{\pi}{\neg}\overset{\pi}{\neg}\tau$ so by MoJ, $\vdash \sigma \vee \tau \Rightarrow \left(\overset{\pi}{\neg}\overset{\pi}{\neg}\sigma \vee \tau\right)$ and $\vdash \left(\tau \vee \overset{\pi}{\neg}\overset{\pi}{\neg}\sigma\right) \Rightarrow \left(\overset{\pi}{\neg}\overset{\pi}{\neg}\tau \vee \overset{\pi}{\neg}\overset{\pi}{\neg}\sigma\right)$ so by ToI, $\vdash (\sigma \vee \tau) \Rightarrow \left(\overset{\pi}{\neg}\overset{\pi}{\neg}\sigma \vee \overset{\pi}{\neg}\overset{\pi}{\neg}\tau\right)$ so by contraposition, $\vdash \overset{\pi}{\neg}\left(\overset{\pi}{\neg}\overset{\pi}{\neg}\sigma \vee \overset{\pi}{\neg}\overset{\pi}{\neg}\tau\right) \Rightarrow \overset{\pi}{\neg}(\sigma \vee \tau)$. Now $\vdash \overset{\pi}{\neg}\sigma \wedge \overset{\pi}{\neg}\tau \Rightarrow \overset{\pi}{\neg}\sigma$ and $\vdash \overset{\pi}{\neg}\sigma \wedge \overset{\pi}{\neg}\tau \Rightarrow \overset{\pi}{\neg}\tau$ so by contrapositing, we have: $\vdash \overset{\pi}{\neg}\overset{\pi}{\neg}\sigma \Rightarrow \overset{\pi}{\neg}\left(\overset{\pi}{\neg}\sigma \wedge \overset{\pi}{\neg}\tau\right)$ and $\vdash \overset{\pi}{\neg}\overset{\pi}{\neg}\tau \Rightarrow \overset{\pi}{\neg}\left(\overset{\pi}{\neg}\sigma \wedge \overset{\pi}{\neg}\tau\right)$ and thus by the previous derived rule of inference, $\vdash \left(\overset{\pi}{\neg}\overset{\pi}{\neg}\sigma \vee \overset{\pi}{\neg}\overset{\pi}{\neg}\tau\right) \Rightarrow \overset{\pi}{\neg}\left(\overset{\pi}{\neg}\sigma \wedge \overset{\pi}{\neg}\tau\right)$ and then contraposition gives:

$$\vdash \overset{\pi}{\neg}\overset{\pi}{\neg}\left(\overset{\pi}{\neg}\sigma \wedge \overset{\pi}{\neg}\tau\right) \Rightarrow \overset{\pi}{\neg}\left(\overset{\pi}{\neg}\overset{\pi}{\neg}\sigma \vee \overset{\pi}{\neg}\overset{\pi}{\neg}\tau\right).$$

But by a previous theorem, $\vdash \overset{\pi}{\neg}\overset{\pi}{\neg}\left(\overset{\pi}{\neg}\sigma \wedge \overset{\pi}{\neg}\tau\right) \Leftrightarrow \left(\overset{\pi}{\neg}\sigma \wedge \overset{\pi}{\neg}\tau\right)$, so putting the implications together by ToI, $\vdash \left(\overset{\pi}{\neg}\sigma \wedge \overset{\pi}{\neg}\tau\right) \Rightarrow \overset{\pi}{\neg}(\sigma \vee \tau)$ which is the right to left implication. □

Lemma 67 *If* $\vdash \varphi$ *then* $\vdash \sigma \Rightarrow (\sigma \wedge \varphi)$.

Proof: $\vdash \varphi \Rightarrow (\sigma \Rightarrow \varphi)$ so by MP, $\vdash \sigma \Rightarrow \varphi$ and $\vdash \sigma \Rightarrow \sigma$ so $\vdash \sigma \Rightarrow (\sigma \wedge \varphi)$. □

Theorem 68 $\vdash \overset{\pi}{\neg}(\sigma \vee \pi) \Leftrightarrow \overset{\pi}{\neg}\sigma$.

Proof: Using weak DeMorgan, $\vdash \overset{\pi}{\neg}(\sigma \vee \pi) \Leftrightarrow \overset{\pi}{\neg}\sigma \wedge \overset{\pi}{\neg}\pi \Rightarrow \overset{\pi}{\neg}\sigma$. Since $\vdash \overset{\pi}{\neg}\pi$ and $\vdash \overset{\pi}{\neg}\sigma \Rightarrow \overset{\pi}{\neg}\sigma \wedge \overset{\pi}{\neg}\pi$ by the lemma, we have the right to left implication. □

Theorem 69 $\vdash (\sigma \vee \pi) \wedge (\sigma \Rightarrow \pi) \Leftrightarrow \pi$.

Proof: Since $\vdash \pi \Rightarrow (\sigma \vee \pi)$ and $\vdash \pi \Rightarrow (\sigma \Rightarrow \pi)$, we have $\vdash \pi \Rightarrow [(\sigma \vee \pi) \wedge (\sigma \Rightarrow \pi)]$ which is the right to left implication. Conversely plug $\sigma \vee \pi$ into modus ponens to get $\vdash (\sigma \vee \pi) \wedge ((\sigma \vee \pi) \Rightarrow \pi) \Rightarrow \pi$. By the previous theorem, $\vdash \overset{\pi}{\neg}\sigma \Rightarrow \overset{\pi}{\neg}(\sigma \vee \pi)$ so by MoM rule of inference $\vdash (\sigma \vee \pi) \wedge \overset{\pi}{\neg}\sigma \Rightarrow (\sigma \vee \pi) \wedge ((\sigma \vee \pi) \Rightarrow \pi)$ so by ToI, we have the left to right implication. □

Theorem 70 $\vdash (\sigma \vee \pi) \Rightarrow \overset{\pi}{\neg}\overset{\pi}{\neg}\sigma$.

Proof: $\vdash \sigma \Rightarrow \overset{\pi}{\neg}\overset{\pi}{\neg}\sigma$ and $\vdash \pi \Rightarrow \overset{\pi}{\neg}\overset{\pi}{\neg}\sigma$ so $\vdash (\sigma \vee \pi) \Rightarrow \overset{\pi}{\neg}\overset{\pi}{\neg}\sigma$. □

We now have some theorems about the π-boundary of σ defined as: $\partial_{\pi}\sigma = \sigma \vee \overset{\pi}{\neg}\sigma$.

Theorem 71 $\vdash \overset{\pi}{\neg}\left(\sigma \vee \overset{\pi}{\neg}\sigma\right) = \overset{\pi}{\neg}\partial_{\pi}\sigma \Leftrightarrow \pi$.

Proof: Using weak DeMorgan, $\vdash \overset{\pi}{\neg}\left(\sigma \vee \overset{\pi}{\neg}\sigma\right) \Leftrightarrow \overset{\pi}{\neg}\sigma \wedge \overset{\pi}{\neg}\overset{\pi}{\neg}\sigma \Leftrightarrow \pi$. □

Theorem 72 $\vdash \overset{\pi}{\neg}\overset{\pi}{\neg}\partial_\pi\sigma$, *i.e.,* $\vdash \overset{\pi}{\neg}\overset{\pi}{\neg}\left(\sigma \vee \overset{\pi}{\neg}\sigma\right)$.

Proof: $\vdash \overset{\pi}{\neg}\pi$ and $\vdash \overset{\pi}{\neg}\pi \Rightarrow \overset{\pi}{\neg}\overset{\pi}{\neg}\partial_\pi\sigma$ so the result follows by MP.   □

Theorem 73 $\partial_\pi (\partial_\pi\sigma) \Leftrightarrow \partial_\pi\sigma$. *(idempotency of π-boundary operation)*

Proof: $\vdash \partial_\pi (\partial_\pi\sigma) = \partial_\pi\sigma \vee \overset{\pi}{\neg}\partial_\pi\sigma \Leftrightarrow \partial_\pi\sigma \vee \pi = \sigma \vee (\sigma \Rightarrow \pi) \vee \pi$. But $\vdash \pi \Rightarrow (\sigma \Rightarrow \pi)$ and $\vdash (\sigma \Rightarrow \pi) \Rightarrow (\sigma \vee (\sigma \Rightarrow \pi))$ so $\vdash \pi \Rightarrow (\sigma \vee (\sigma \Rightarrow \pi))$ and thus using MoJ and absorption, $\vdash \sigma \vee (\sigma \Rightarrow \pi \vee \pi) \Rightarrow (\sigma \vee (\sigma \Rightarrow \pi))$ so we have the left to right implication. Conversely, $\vdash \partial_\pi\sigma \Rightarrow \partial_\pi\sigma \vee \pi$ so by the previous result, the right to left implication also holds.   □

Theorem 74 *If $\vdash \pi \Rightarrow \sigma$ then the following conditions are equivalent: 1)* $\vdash \overset{\pi}{\neg}\sigma \Leftrightarrow \pi$; *2)* $\vdash \partial_\pi\sigma \Leftrightarrow \sigma$, *and 3)* $\vdash \overset{\pi}{\neg}\overset{\pi}{\neg}\sigma$.

Proof: Assuming $\vdash \overset{\pi}{\neg}\sigma \Leftrightarrow \pi$ so that by MoJ, $\vdash \left(\sigma \vee \overset{\pi}{\neg}\sigma\right) \Leftrightarrow (\sigma \vee \pi)$ but since $\vdash \pi \Rightarrow \sigma$, $\vdash (\sigma \vee \pi) \Leftrightarrow \sigma$ so 2) follows. Conversely, if $\vdash \left(\sigma \vee \overset{\pi}{\neg}\sigma\right) \Leftrightarrow \sigma$, then by contraposition, $\vdash \overset{\pi}{\neg}\sigma \Leftrightarrow \overset{\pi}{\neg}\left(\sigma \vee \overset{\pi}{\neg}\sigma\right)$ where by a previous theorem, we have $\vdash \overset{\pi}{\neg}\left(\sigma \vee \overset{\pi}{\neg}\sigma\right) \Leftrightarrow \pi$ so 1) and 2) are equivalent. If 1) holds, then $\vdash \overset{\pi}{\neg}\overset{\pi}{\neg}\sigma \Leftrightarrow \overset{\pi}{\neg}\pi$ so 3) follows by MP. If 3) holds, then $\vdash \overset{\pi}{\neg}\overset{\pi}{\neg}\sigma \Rightarrow \left(1 \Rightarrow \overset{\pi}{\neg}\overset{\pi}{\neg}\sigma\right)$ yields $\vdash 1 \Rightarrow \overset{\pi}{\neg}\overset{\pi}{\neg}\sigma$ so contraposition yields: $\vdash \overset{\pi}{\neg}\overset{\pi}{\neg}\overset{\pi}{\neg}\sigma \Rightarrow \overset{\pi}{\neg}1$ where $\vdash \overset{\pi}{\neg}\sigma \Rightarrow \overset{\pi}{\neg}\overset{\pi}{\neg}\overset{\pi}{\neg}\sigma$ and $\vdash \overset{\pi}{\neg}1 \Rightarrow \pi$ so $\vdash \overset{\pi}{\neg}\sigma \Rightarrow \pi$ and the reverse implication is $\vdash \pi \Rightarrow (\sigma \Rightarrow \pi)$ so 1) follows.   □

Theorem 75 *If a π-regular element $\overset{\pi}{\neg}\sigma$ is also a π-boundary, then $\vdash \overset{\pi}{\neg}\sigma$.*

Proof: If it is a π-boundary, then $\vdash \overset{\pi}{\neg}\overset{\pi}{\neg}\left(\overset{\pi}{\neg}\sigma\right)$ but $\vdash \overset{\pi}{\neg}\overset{\pi}{\neg}\overset{\pi}{\neg}\sigma \Rightarrow \overset{\pi}{\neg}\sigma$ so $\vdash \overset{\pi}{\neg}\sigma$.   □

Theorem 76 $\vdash (\sigma \vee \tau) \Rightarrow \left(\overset{\pi}{\neg}\sigma \Rightarrow \overset{\pi}{\neg}\overset{\pi}{\neg}\tau\right)$.

Proof: $\vdash \tau \Rightarrow \overset{\pi}{\neg}\overset{\pi}{\neg}\tau$ and $\vdash \overset{\pi}{\neg}\overset{\pi}{\neg}\tau \Rightarrow \left(\overset{\pi}{\neg}\sigma \Rightarrow \overset{\pi}{\neg}\overset{\pi}{\neg}\tau\right)$ so $\vdash \tau \Rightarrow \left(\overset{\pi}{\neg}\sigma \Rightarrow \overset{\pi}{\neg}\overset{\pi}{\neg}\tau\right)$. Using the equivalence of contraposition in the Boolean core L_π, $\vdash \left(\overset{\pi}{\neg}\sigma \Rightarrow \overset{\pi}{\neg}\overset{\pi}{\neg}\tau\right) \Leftrightarrow \left(\overset{\pi}{\neg}\tau \Rightarrow \overset{\pi}{\neg}\overset{\pi}{\neg}\sigma\right)$ so the same argument shows that $\vdash \sigma \Rightarrow \left(\overset{\pi}{\neg}\sigma \Rightarrow \overset{\pi}{\neg}\overset{\pi}{\neg}\tau\right)$ and thus using the greatest lower bound property of the join, $\vdash (\sigma \vee \tau) \Rightarrow \left(\overset{\pi}{\neg}\sigma \Rightarrow \overset{\pi}{\neg}\overset{\pi}{\neg}\tau\right)$.
□

Corollary 18 $\vdash (\sigma \vee \pi) \Rightarrow \overset{\pi}{\neg}\overset{\pi}{\neg}\sigma$.

Proof: $\vdash (\sigma \vee \pi) \Rightarrow \left(\overset{\pi}{\neg}\sigma \Rightarrow \overset{\pi}{\neg}\overset{\pi}{\neg}\pi \right) \Leftrightarrow \left(\overset{\pi}{\neg}\sigma \Rightarrow \pi \right) = \overset{\pi}{\neg}\overset{\pi}{\neg}\sigma$. □

Theorem 77 $\vdash (\sigma \Rightarrow (\sigma \Rightarrow \pi)) \Leftrightarrow (\sigma \Rightarrow \pi)$, *i.e.*, $\vdash \left(\sigma \Rightarrow \overset{\pi}{\neg}\sigma \right) \Leftrightarrow \overset{\pi}{\neg}\sigma$. *(reductio)*

Proof: $\vdash (\sigma \Rightarrow \pi) \Rightarrow (\sigma \Rightarrow (\sigma \Rightarrow \pi))$ so it remains to show the left to right implication. Now the previous theorem $\vdash ((\sigma \vee \pi) \Rightarrow \pi) \Leftrightarrow (\sigma \Rightarrow \pi)$ gives $\vdash \left(\sigma \Rightarrow \overset{\pi}{\neg}\sigma \right) \Leftrightarrow \left(\left(\sigma \vee \overset{\pi}{\neg}\sigma \right) \Rightarrow \overset{\pi}{\neg}\sigma \right)$ and by a previous theorem $\vdash \left(\tau \Rightarrow \overset{\pi}{\neg}\sigma \right) \Leftrightarrow \left(\overset{\pi}{\neg}\overset{\pi}{\neg}\tau \Rightarrow \overset{\pi}{\neg}\sigma \right)$ so that putting it together:

$$\vdash \left(\sigma \Rightarrow \overset{\pi}{\neg}\sigma \right) \Rightarrow \left[\overset{\pi}{\neg}\overset{\pi}{\neg} \left(\sigma \vee \overset{\pi}{\neg}\sigma \right) \Rightarrow \overset{\pi}{\neg}\sigma \right].$$

But $\vdash \overset{\pi}{\neg}\overset{\pi}{\neg} \left(\sigma \vee \overset{\pi}{\neg}\sigma \right)$ so $\vdash \left(\sigma \Rightarrow \overset{\pi}{\neg}\sigma \right) \Rightarrow \overset{\pi}{\neg}\sigma$. □

Theorem 78 $\vdash \left(\tau \Rightarrow \overset{\pi}{\neg}\sigma \right) \Rightarrow \left[(\tau \Rightarrow \sigma) \Rightarrow \overset{\pi}{\neg}\tau \right]$.

Proof: $\vdash \left(\tau \Rightarrow \overset{\pi}{\neg}\sigma \right) \Leftrightarrow \left(\sigma \Rightarrow \overset{\pi}{\neg}\tau \right)$ by interchange, and $\vdash (\tau \Rightarrow \sigma) \Rightarrow \left[\left(\sigma \Rightarrow \overset{\pi}{\neg}\tau \right) \Rightarrow \left(\tau \Rightarrow \overset{\pi}{\neg}\tau \right) \right]$ by ToI. But by reductio, $\vdash \left(\tau \Rightarrow \overset{\pi}{\neg}\tau \right) \Rightarrow \overset{\pi}{\neg}\tau$ so $\vdash \left(\left(\sigma \Rightarrow \overset{\pi}{\neg}\tau \right) \Rightarrow \left(\tau \Rightarrow \overset{\pi}{\neg}\tau \right) \right) \Rightarrow \left(\left(\sigma \Rightarrow \overset{\pi}{\neg}\tau \right) \Rightarrow \overset{\pi}{\neg}\tau \right)$ and thus by ToI, $\vdash (\tau \Rightarrow \sigma) \Rightarrow \left[\left(\sigma \Rightarrow \overset{\pi}{\neg}\tau \right) \Rightarrow \overset{\pi}{\neg}\tau \right]$ so by interchange, $\vdash \left(\sigma \Rightarrow \overset{\pi}{\neg}\tau \right) \Rightarrow \left[(\tau \Rightarrow \sigma) \Rightarrow \overset{\pi}{\neg}\tau \right]$ and by interchange, $\vdash \left(\tau \Rightarrow \overset{\pi}{\neg}\sigma \right) \Rightarrow \left((\tau \Rightarrow \sigma) \Rightarrow \overset{\pi}{\neg}\tau \right)$. □

Corollary 19 $\vdash \left(\tau \Rightarrow \overset{\pi}{\neg}\sigma \right) \Leftrightarrow \left[(\tau \Rightarrow \sigma) \Rightarrow \overset{\pi}{\neg}\tau \right]$. *[86, Formal 18, p. 25]*

Proof: $\vdash \sigma \Rightarrow (\tau \Rightarrow \sigma)$ and then

$$\vdash (\sigma \Rightarrow (\tau \Rightarrow \sigma)) \Rightarrow (((\tau \Rightarrow \sigma) \Rightarrow (\tau \Rightarrow \pi)) \Rightarrow (\sigma \Rightarrow (\tau \Rightarrow \pi)))$$

so that $\vdash ((\tau \Rightarrow \sigma) \Rightarrow (\tau \Rightarrow \pi)) \Rightarrow (\sigma \Rightarrow (\tau \Rightarrow \pi))$ and then the result follows by interchange in the consequent. □

Corollary 20 $\vdash (\pi \Rightarrow \sigma) \Rightarrow ((\sigma \Rightarrow \pi) \Rightarrow \sigma) \Leftrightarrow (\sigma \Rightarrow \pi) \Rightarrow ((\pi \Rightarrow \sigma) \Rightarrow \pi)$. *[86, Formula 19, p. 25]*

Proof: $\vdash \sigma \Rightarrow ((\sigma \Rightarrow \pi) \Rightarrow \pi)$ and then using the covariance of $(\sigma \Rightarrow \pi) \Rightarrow -$ which means plugging into ToI and using the double negation theorem and MP to get: $\vdash ((\sigma \Rightarrow \pi) \Rightarrow \sigma) \Rightarrow ((\sigma \Rightarrow \pi) \Rightarrow ((\sigma \Rightarrow \pi) \Rightarrow \pi))$ and then do it again with $(\pi \Rightarrow \sigma) \Rightarrow -$ applied to the whole formula to get:

$$\vdash [(\pi \Rightarrow \sigma) \Rightarrow ((\sigma \Rightarrow \pi) \Rightarrow \sigma)] \Rightarrow [(\pi \Rightarrow \sigma) \Rightarrow ((\sigma \Rightarrow \pi) \Rightarrow ((\sigma \Rightarrow \pi) \Rightarrow \pi))]$$

and then reductio and interchange are applied to the consequent to get the result. The opposite implication is obtained by symmetry. $\qquad \square$

If we analyze the partition interpretation of the formula, we see that one side is π-regular and the other side is σ-regular, and they are equal because the one side is the B's that equal a C, and the other side is the C's that equal a B.

Corollary 21 $\vdash \left[(\sigma \Rightarrow \pi) \Rightarrow \overset{\sigma\ \sigma}{\neg\neg}\pi \right] \Leftrightarrow \left[(\pi \Rightarrow \sigma) \Rightarrow \overset{\pi\ \pi}{\neg\neg}\sigma \right].$ $\qquad \square$

Corollary 22 $\vdash \left((\sigma \vee \pi) \Rightarrow \overset{\pi}{\neg}\tau \right) \Leftrightarrow \left(\sigma \Rightarrow \overset{\pi}{\neg}\tau \right).$

Proof: By weak DeMorgan, $\vdash \left((\sigma \vee \pi) \Rightarrow \overset{\pi}{\neg}\tau \right) \Leftrightarrow \left(\left(\sigma \Rightarrow \overset{\pi}{\neg}\tau \right) \wedge \left(\pi \Rightarrow \overset{\pi}{\neg}\tau \right) \right)$ and $\vdash \pi \Rightarrow \overset{\pi}{\neg}\tau$ so by a previous result, the corollary holds. $\qquad \square$

Theorem 79 $\vdash \quad (\sigma \Rightarrow \tau) \quad \Rightarrow \quad ((\varphi \Rightarrow \pi) \Rightarrow ((\sigma \vee \varphi) \Rightarrow (\tau \vee \pi))).$ *(strong monotony for join)*

Proof: By weak monotony of join, $\vdash (\sigma \Rightarrow \tau) \Rightarrow ((\sigma \vee \varphi) \Rightarrow (\tau \vee \varphi))$ and $\vdash (\varphi \Rightarrow \pi) \Rightarrow ((\tau \vee \varphi) \Rightarrow (\tau \vee \pi))$. By ToI, we have:

$$\vdash ((\sigma \vee \varphi) \Rightarrow (\tau \vee \varphi)) \Rightarrow (((\tau \vee \varphi) \Rightarrow (\tau \vee \pi)) \Rightarrow ((\sigma \vee \varphi) \Rightarrow (\tau \vee \pi)))$$

and thus: $\vdash (\sigma \Rightarrow \tau) \Rightarrow (((\tau \vee \varphi) \Rightarrow (\tau \vee \pi)) \Rightarrow ((\sigma \vee \varphi) \Rightarrow (\tau \vee \pi)))$, and by interchange,

$$\vdash ((\tau \vee \varphi) \Rightarrow (\tau \vee \pi)) \Rightarrow [(\sigma \Rightarrow \tau) \Rightarrow ((\sigma \vee \varphi) \Rightarrow (\tau \vee \pi))],$$

and thus: $\vdash (\varphi \Rightarrow \pi) \Rightarrow [(\sigma \Rightarrow \tau) \Rightarrow ((\sigma \vee \varphi) \Rightarrow (\tau \vee \pi))]$ which is interchanged to give the result. $\qquad \square$

So far all the results are intuitionistically valid since all the axioms and rules of inference are intuitionistically valid. Now we introduce an intuitionistically invalid but partitionally valid axiom.

Axiom 4 (Join = Join in Boolean core) $\vdash \overset{\pi\ \pi}{\neg\neg} \left(\overset{\pi}{\neg}\sigma \vee \overset{\pi}{\neg}\tau \right) \Rightarrow \left(\overset{\pi}{\neg}\sigma \vee \overset{\pi}{\neg}\tau \right).$

Corollary 23 $\vdash \overset{\pi\pi}{\neg\neg}\left(\overset{\pi}{\neg}\sigma \vee \overset{\pi}{\neg}\tau\right) \Leftrightarrow \left(\overset{\pi}{\neg}\sigma \vee \overset{\pi}{\neg}\tau\right).$

Proof: The converse is just double π-negation so we have the bi-implication, which says that the join of two π-regular elements is π-regular. □

Theorem 80 $\vdash \overset{\pi\pi}{\neg\neg}(\sigma \vee \tau) \Leftrightarrow \left(\overset{\pi\pi}{\neg\neg}\sigma \vee \overset{\pi\pi}{\neg\neg}\tau\right).$

Proof: By weak DeMorgan, LHS $\Leftrightarrow \overset{\pi}{\neg}\left(\overset{\pi}{\neg}\sigma \wedge \overset{\pi}{\neg}\tau\right)$. Using the axiom and corollary, $\overset{\pi\pi}{\neg\neg}\left(\overset{\pi\pi}{\neg\neg}\sigma \vee \overset{\pi\pi}{\neg\neg}\tau\right) \Leftrightarrow \left(\overset{\pi\pi}{\neg\neg}\sigma \vee \overset{\pi\pi}{\neg\neg}\tau\right) =$ RHS and by weak DeMorgan and triple negation, $\vdash \overset{\pi\pi}{\neg\neg}\left(\overset{\pi\pi}{\neg\neg}\sigma \vee \overset{\pi\pi}{\neg\neg}\tau\right) \Leftrightarrow \overset{\pi}{\neg}\left(\overset{\pi\pi\pi}{\neg\neg\neg}\sigma \wedge \overset{\pi\pi\pi}{\neg\neg\neg}\tau\right) \Leftrightarrow \overset{\pi}{\neg}\left(\overset{\pi}{\neg}\sigma \wedge \overset{\pi}{\neg}\tau\right)$ so the result follows. □

Theorem 81 $\vdash \left(\overset{\pi\pi}{\neg\neg}\sigma \vee \overset{\pi}{\neg}\sigma\right)$ *(weak law of excluded middle = excluded middle in Boolean core L_π)*

Proof: By the previous theorem with $\tau = \overset{\pi}{\neg}\sigma$, we have: $\vdash \overset{\pi\pi}{\neg\neg}\left(\sigma \vee \overset{\pi}{\neg}\sigma\right) \Leftrightarrow \left(\overset{\pi\pi}{\neg\neg}\sigma \vee \overset{\pi\pi\pi}{\neg\neg\neg}\sigma\right) \Leftrightarrow \left(\overset{\pi\pi}{\neg\neg}\sigma \vee \overset{\pi}{\neg}\sigma\right)$ so the result follows by the previous theorem $\vdash \overset{\pi\pi}{\neg\neg}\left(\sigma \vee \overset{\pi}{\neg}\sigma\right)$. □

Theorem 82 $\vdash \left(\tau \Rightarrow \overset{\pi}{\neg}\sigma\right) \Leftrightarrow \left(\overset{\pi}{\neg}\tau \vee \overset{\pi}{\neg}\sigma\right).$ *"disjunctive implication")*

Proof: $\vdash \left(\tau \Rightarrow \overset{\pi}{\neg}\sigma\right) \Leftrightarrow \left(\sigma \Rightarrow \overset{\pi}{\neg}\tau\right)$ and $\vdash \left(\sigma \Rightarrow \overset{\pi}{\neg}\tau\right) \Rightarrow \left(\left(\sigma \vee \overset{\pi}{\neg}\sigma\right) \Rightarrow \left(\overset{\pi}{\neg}\tau \vee \overset{\pi}{\neg}\sigma\right)\right)$ and thus by the corollary, $\vdash \left(\sigma \Rightarrow \overset{\pi}{\neg}\tau\right) \Rightarrow \left(\left(\sigma \vee \overset{\pi}{\neg}\sigma\right) \Rightarrow \overset{\pi\pi}{\neg\neg}\left(\overset{\pi}{\neg}\tau \vee \overset{\pi}{\neg}\sigma\right)\right)$. Then using $\vdash \left(\varphi \Rightarrow \overset{\pi}{\neg}\phi\right) \Leftrightarrow \left(\overset{\pi\pi}{\neg\neg}\varphi \Rightarrow \overset{\pi}{\neg}\phi\right)$, we have

$$\vdash \left(\sigma \Rightarrow \overset{\pi}{\neg}\tau\right) \Rightarrow \left(\overset{\pi\pi}{\neg\neg}\left(\sigma \vee \overset{\pi}{\neg}\sigma\right) \Rightarrow \overset{\pi\pi}{\neg\neg}\left(\overset{\pi}{\neg}\tau \vee \overset{\pi}{\neg}\sigma\right)\right)$$

but $\vdash \overset{\pi\pi}{\neg\neg}\left(\sigma \vee \overset{\pi}{\neg}\sigma\right)$ so that $\vdash \left(\sigma \Rightarrow \overset{\pi}{\neg}\tau\right) \Rightarrow \overset{\pi\pi}{\neg\neg}\left(\overset{\pi}{\neg}\tau \vee \overset{\pi}{\neg}\sigma\right)$ and thus $\left(\tau \Rightarrow \overset{\pi}{\neg}\sigma\right) \Rightarrow \left(\overset{\pi}{\neg}\tau \vee \overset{\pi}{\neg}\sigma\right)$. Conversely, $\vdash \sigma \Rightarrow \left(\overset{\pi}{\neg}\sigma \Rightarrow \overset{\pi}{\neg}\tau\right)$ and $\vdash \left(\left(\overset{\pi}{\neg}\sigma \vee \overset{\pi}{\neg}\tau\right) \Rightarrow \overset{\pi}{\neg}\tau\right) \Leftrightarrow \left(\overset{\pi}{\neg}\sigma \Rightarrow \overset{\pi}{\neg}\tau\right)$ so that $\vdash \sigma \Rightarrow \left(\left(\overset{\pi}{\neg}\sigma \vee \overset{\pi}{\neg}\tau\right) \Rightarrow \overset{\pi}{\neg}\tau\right)$ and the right to left implication follows by interchange. □

Theorem 83 $\vdash \left(\overset{\pi}{\neg}\sigma \vee \tau\right) \Leftrightarrow \left(\overset{\pi\pi}{\neg\neg}\sigma \Rightarrow (\tau \vee \pi)\right).$ *(implication-join theorem)*

Proof: $\vdash \left(\overset{\pi}{\neg}\overset{\pi}{\neg}\sigma \Rightarrow \tau \right) \vee \left(\overset{\pi}{\neg}\overset{\pi}{\neg}\sigma \Rightarrow \pi \right) \Rightarrow \left(\overset{\pi}{\neg}\overset{\pi}{\neg}\sigma \Rightarrow (\tau \vee \pi) \right.$ and $\vdash \tau \Rightarrow \left(\overset{\pi}{\neg}\overset{\pi}{\neg}\sigma \Rightarrow \tau \right)$ and $\vdash \left(\overset{\pi}{\neg}\overset{\pi}{\neg}\sigma \Rightarrow \pi \right) \Leftrightarrow \overset{\pi}{\neg}\sigma$ so $\vdash \left(\overset{\pi}{\neg}\sigma \vee \tau \right) \Rightarrow \left(\overset{\pi}{\neg}\overset{\pi}{\neg}\sigma \Rightarrow \tau \right) \vee \left(\overset{\pi}{\neg}\overset{\pi}{\neg}\sigma \Rightarrow \pi \right)$ and thus the left to right implication holds. For the other direction, weak monotony of join gives:

$$\vdash \left(\overset{\pi}{\neg}\overset{\pi}{\neg}\sigma \Rightarrow (\tau \vee \pi) \right) \Rightarrow \left(\overset{\pi}{\neg}\overset{\pi}{\neg}\sigma \vee \overset{\pi}{\neg}\sigma \right) \Rightarrow \left(\tau \vee \pi \vee \overset{\pi}{\neg}\sigma \right)$$

and since $\overset{\pi}{\neg}\overset{\pi}{\neg}\sigma \vee \overset{\pi}{\neg}\sigma$ is a theorem, we have: $\vdash \left(\overset{\pi}{\neg}\overset{\pi}{\neg}\sigma \Rightarrow (\tau \vee \pi) \right) \Rightarrow \left(\tau \vee \pi \vee \overset{\pi}{\neg}\sigma \right)$ and $\vdash \pi \Rightarrow \left(\overset{\pi}{\neg}\sigma \right)$ so that $\vdash \left(\pi \vee \overset{\pi}{\neg}\sigma \right) \Leftrightarrow \overset{\pi}{\neg}\sigma$ and thus the right to left implication holds. $\square$

Note that the disjunction implication theorem also follows as the special case where τ is replaced by $\overset{\pi}{\neg}\tau$. The excluded middle principle in the Boolean core L_π also follows by taking τ as $\overset{\pi}{\neg}\overset{\pi}{\neg}\sigma$. The following result also gives the π-boundary as an implication.

Corollary 24 $\vdash \partial_\pi \sigma = (\sigma \vee \overset{\pi}{\neg}\sigma) \Leftrightarrow \left(\overset{\pi}{\neg}\overset{\pi}{\neg}\sigma \Rightarrow (\sigma \vee \pi) \right).$ $\qquad\qquad$ $\square$

Theorem 84 $\vdash \varphi \vee \left(\overset{\pi}{\neg}\sigma \wedge \overset{\pi}{\neg}\tau \right) \Leftrightarrow (\varphi \vee \overset{\pi}{\neg}\sigma) \wedge \left(\varphi \vee \overset{\pi}{\neg}\tau \right).$ *(Ore's associability theorem)*

Proof: By weak DeMorgan and triple negation, $\vdash \left(\overset{\pi}{\neg}\sigma \wedge \overset{\pi}{\neg}\tau \right) \Leftrightarrow \overset{\pi}{\neg}\left(\overset{\pi}{\neg}\overset{\pi}{\neg}\sigma \vee \overset{\pi}{\neg}\overset{\pi}{\neg}\tau \right)$ so the LHS is $\varphi \vee \overset{\pi}{\neg}\left(\overset{\pi}{\neg}\overset{\pi}{\neg}\sigma \vee \overset{\pi}{\neg}\overset{\pi}{\neg}\tau \right)$ so the implication-join theorem gives:

$$\vdash \varphi \vee \overset{\pi}{\neg}\left(\overset{\pi}{\neg}\overset{\pi}{\neg}\sigma \vee \overset{\pi}{\neg}\overset{\pi}{\neg}\tau \right) \Leftrightarrow \left(\overset{\pi}{\neg}\overset{\pi}{\neg}\left(\overset{\pi}{\neg}\overset{\pi}{\neg}\sigma \vee \overset{\pi}{\neg}\overset{\pi}{\neg}\tau \right) \Rightarrow (\varphi \vee \pi) \right),$$

and then by the unified join theorem,

$$\vdash \varphi \vee \left(\overset{\pi}{\neg}\sigma \wedge \overset{\pi}{\neg}\tau \right) \Leftrightarrow \left(\left(\overset{\pi}{\neg}\overset{\pi}{\neg}\sigma \vee \overset{\pi}{\neg}\overset{\pi}{\neg}\tau \right) \Rightarrow (\varphi \vee \pi) \right).$$

Then using weak DeMorgan with $(\varphi \vee \pi)$-negation, we have:

$$\vdash \varphi \vee \left(\overset{\pi}{\neg}\sigma \wedge \overset{\pi}{\neg}\tau \right) \Leftrightarrow \left(\overset{\pi}{\neg}\overset{\pi}{\neg}\sigma \Rightarrow (\varphi \vee \pi) \right) \wedge \left(\overset{\pi}{\neg}\overset{\pi}{\neg}\tau \Rightarrow (\varphi \vee \pi) \right)$$

and then by the implication-join theorem again,

$$\vdash\vdash \varphi \vee \left(\overset{\pi}{\neg}\sigma \wedge \overset{\pi}{\neg}\tau \right) \Leftrightarrow (\varphi \vee \overset{\pi}{\neg}\sigma) \wedge \left(\varphi \vee \overset{\pi}{\neg}\tau \right).$$ $\qquad\qquad$ $\square$

Theorem 85 $\vdash \left[\left(\sigma \vee \overset{\pi}{\neg}\sigma\right) \wedge \overset{\pi\,\pi}{\neg\neg}\sigma\right] \Leftrightarrow (\sigma \vee \pi)$ *(π-relative Core + boundary law)*

Proof: By the previous theorem for the core and MP,

$$\vdash \left[\left(\sigma \vee \overset{\pi}{\neg}\sigma\right) \wedge \overset{\pi\,\pi}{\neg\neg}\sigma\right] \Leftrightarrow \left(\overset{\pi\,\pi}{\neg\neg}\sigma \Rightarrow (\sigma \vee \pi)\right) \wedge \overset{\pi\,\pi}{\neg\neg}\sigma \Leftrightarrow (\sigma \vee \pi) \wedge \overset{\pi\,\pi}{\neg\neg}\sigma.$$

By a previous theorem, $\vdash (\sigma \vee \pi) \Rightarrow \overset{\pi\,\pi}{\neg\neg}\sigma$ so that $\vdash \left((\sigma \vee \pi) \wedge \overset{\pi\,\pi}{\neg\neg}\sigma\right) \Leftrightarrow (\sigma \vee \pi)$. $\qquad\square$

Theorem 86 $\vdash \partial_\pi (\sigma \vee \tau) \Leftrightarrow (\partial_\pi \sigma \vee \tau) \wedge (\sigma \vee \partial_\pi \tau)$. *(Leibniz's rule)*

Proof: $\vdash \partial_\pi (\sigma \vee \tau) \Leftrightarrow (\sigma \vee \tau) \vee \overset{\pi}{\neg}(\sigma \vee \tau) \Leftrightarrow (\sigma \vee \tau) \vee \left(\overset{\pi}{\neg}\sigma \wedge \overset{\pi}{\neg}\tau\right)$ so by Ore's theorem,

$$\vdash \partial_\pi (\sigma \vee \tau) \Leftrightarrow \left(\sigma \vee \tau \vee \overset{\pi}{\neg}\sigma\right) \wedge \left(\sigma \vee \tau \vee \overset{\pi}{\neg}\tau\right) \Leftrightarrow (\partial_\pi \sigma \vee \tau) \wedge (\sigma \vee \partial_\pi \tau). \qquad\square$$

Theorem 87 $\vdash \varphi \Rightarrow \left(\overset{\pi}{\neg}\sigma \wedge \overset{\pi}{\neg}\tau\right) \Leftrightarrow \left(\varphi \Rightarrow \overset{\pi}{\neg}\sigma\right) \wedge \left(\varphi \Rightarrow \overset{\pi}{\neg}\tau\right)$. *(implication to π-regular meet = meet of implications)*

Proof: $\vdash \varphi \Rightarrow \left(\overset{\pi}{\neg}\sigma \wedge \overset{\pi}{\neg}\tau\right) \Leftrightarrow \overset{\pi}{\neg}\varphi \vee \left(\overset{\pi}{\neg}\sigma \wedge \overset{\pi}{\neg}\tau\right) \Leftrightarrow \left(\overset{\pi}{\neg}\varphi \vee \overset{\pi}{\neg}\sigma\right) \wedge \left(\overset{\pi}{\neg}\varphi \vee \overset{\pi}{\neg}\tau\right)$ so the result follows by disjunctive implication again. $\qquad\square$

Theorem 88 $\vdash \left(\overset{\pi}{\neg}\sigma \vee \overset{\pi}{\neg}\tau\right) \Rightarrow \overset{\pi}{\neg}(\sigma \wedge \tau)$. *(half of strong DeMorgan)*

Proof: $\vdash \overset{\pi\,\pi}{\neg\neg}\left(\overset{\pi}{\neg}\sigma \vee \overset{\pi}{\neg}\tau\right) \Leftrightarrow \left(\overset{\pi}{\neg}\sigma \vee \overset{\pi}{\neg}\tau\right) \Leftrightarrow \overset{\pi}{\neg}\left(\overset{\pi\,\pi}{\neg\neg}\sigma \wedge \overset{\pi\,\pi}{\neg\neg}\tau\right)$. Now $\vdash (\sigma \wedge \tau) \Rightarrow \left(\overset{\pi\,\pi}{\neg\neg}\sigma \wedge \overset{\pi\,\pi}{\neg\neg}\tau\right)$ so contraposition and ToI gives the result. $\qquad\square$

Corollary 25 $\vdash (\sigma \vee \tau) \Rightarrow \overset{\pi}{\neg}\left(\overset{\pi}{\neg}\sigma \wedge \overset{\pi}{\neg}\tau\right)$.

Proof: Contraposite the theorem and use double negation. $\qquad\square$

Corollary 26 $\vdash \left(\sigma \wedge \overset{\pi}{\neg}\tau\right) \Rightarrow \overset{\pi}{\neg}(\sigma \Rightarrow \tau)$.

Proof: $\vdash (\sigma \Rightarrow \tau) \Rightarrow \left(\sigma \Rightarrow \overset{\pi\,\pi}{\neg\neg}\tau\right) \Leftrightarrow \left(\overset{\pi}{\neg}\sigma \vee \overset{\pi\,\pi}{\neg\neg}\tau\right) \Leftrightarrow \overset{\pi}{\neg}\left(\overset{\pi\,\pi}{\neg\neg}\sigma \wedge \overset{\pi}{\neg}\tau\right)$ so contrapositing gives $\vdash \overset{\pi\,\pi}{\neg\neg}\left(\overset{\pi\,\pi}{\neg\neg}\sigma \wedge \overset{\pi}{\neg}\tau\right) \Rightarrow \overset{\pi}{\neg}(\sigma \Rightarrow \tau)$ and the double negation can be deleted and $\vdash (\sigma \wedge \tau) \Rightarrow \left(\overset{\pi\,\pi}{\neg\neg}\sigma \wedge \overset{\pi}{\neg}\tau\right)$ so the result follows. $\qquad\square$

Corollary 27 $\vdash \left(\sigma \Rightarrow \overset{\pi}{\neg}\tau \right) \Rightarrow \overset{\pi}{\neg}(\sigma \wedge \tau).$

Proof: $\vdash \left(\sigma \Rightarrow \overset{\pi}{\neg}\tau \right) \Leftrightarrow \overset{\pi}{\neg}\sigma \vee \overset{\pi}{\neg}\tau \Rightarrow \overset{\pi}{\neg}(\sigma \wedge \tau).$ $\square$

Corollary 28 $\vdash \overset{\pi}{\neg}\overset{\pi}{\neg}(\tau \Rightarrow \sigma) \Rightarrow \left(\overset{\pi}{\neg}\overset{\pi}{\neg}\tau \Rightarrow \overset{\pi}{\neg}\overset{\pi}{\neg}\sigma \right).$

Proof: $\vdash \sigma \Rightarrow \overset{\pi}{\neg}\overset{\pi}{\neg}\sigma$ so that $\vdash (\tau \Rightarrow \sigma) \Rightarrow \left(\tau \Rightarrow \overset{\pi}{\neg}\overset{\pi}{\neg}\sigma \right) \Leftrightarrow \overset{\pi}{\neg}\overset{\pi}{\neg}\tau \Rightarrow \overset{\pi}{\neg}\overset{\pi}{\neg}\sigma$ so that applying double negation and using the fact that the implication of π-regular elements is π-regular so it is equivalent to its own double negation, the result follows. $\square$

Corollary 29 $\vdash \left(\overset{\pi}{\neg}\sigma \Rightarrow \overset{\pi}{\neg}\tau \right) \vee \left(\overset{\pi}{\neg}\tau \Rightarrow \overset{\pi}{\neg}\sigma \right).$

Proof: $\vdash \left(\overset{\pi}{\neg}\sigma \Rightarrow \overset{\pi}{\neg}\tau \right) \vee \left(\overset{\pi}{\neg}\tau \Rightarrow \overset{\pi}{\neg}\sigma \right) \Leftrightarrow \left(\overset{\pi}{\neg}\overset{\pi}{\neg}\sigma \vee \overset{\pi}{\neg}\tau \vee \overset{\pi}{\neg}\overset{\pi}{\neg}\tau\overset{\pi}{\neg}\sigma \right) \Leftrightarrow \left(\overset{\pi}{\neg}\overset{\pi}{\neg}\sigma \vee \overset{\pi}{\neg}\sigma \right) \vee \left(\overset{\pi}{\neg}\overset{\pi}{\neg}\tau \vee \overset{\pi}{\neg}\tau \right)$ so the result follows. $\square$

Corollary 30 $\vdash \overset{\pi}{\neg}\left(\overset{\pi}{\neg}\sigma \wedge \overset{\pi}{\neg}\tau \right) \Leftrightarrow \left(\overset{\pi}{\neg}\overset{\pi}{\neg}\sigma \vee \overset{\pi}{\neg}\overset{\pi}{\neg}\tau \right)$ and $\vdash \left(\overset{\pi}{\neg}\sigma \vee \overset{\pi}{\neg}\tau \right) \Leftrightarrow \overset{\pi}{\neg}\left(\overset{\pi}{\neg}\overset{\pi}{\neg}\sigma \wedge \overset{\pi}{\neg}\overset{\pi}{\neg}\tau \right).$ *(strong DeMorgan in the Boolean core L_π)*

Proof: $\vdash \overset{\pi}{\neg}\overset{\pi}{\neg}(\sigma \vee \tau) \Leftrightarrow \overset{\pi}{\neg}\left(\overset{\pi}{\neg}\sigma \wedge \overset{\pi}{\neg}\tau \right)$ where $\vdash \left(\overset{\pi}{\neg}\overset{\pi}{\neg}\sigma \vee \overset{\pi}{\neg}\overset{\pi}{\neg}\tau \right) \Rightarrow \overset{\pi}{\neg}\left(\overset{\pi}{\neg}\sigma \wedge \overset{\pi}{\neg}\tau \right)$ from the previous half of strong DeMorgan.

By weak DeMorgan, $\vdash \overset{\pi}{\neg}\left(\overset{\pi}{\neg}\sigma \wedge \overset{\pi}{\neg}\tau \right) \Leftrightarrow \overset{\pi}{\neg}\overset{\pi}{\neg}(\sigma \vee \tau) \Leftrightarrow \overset{\pi}{\neg}\overset{\pi}{\neg}\sigma \vee \overset{\pi}{\neg}\overset{\pi}{\neg}\tau$ so the first result holds. $\vdash \left(\overset{\pi}{\neg}\sigma \vee \overset{\pi}{\neg}\tau \right) \Leftrightarrow \overset{\pi}{\neg}\overset{\pi}{\neg}\left(\overset{\pi}{\neg}\sigma \vee \overset{\pi}{\neg}\tau \right) \Leftrightarrow \overset{\pi}{\neg}\left(\overset{\pi}{\neg}\overset{\pi}{\neg}\sigma \wedge \overset{\pi}{\neg}\overset{\pi}{\neg}\tau \right)$ which is the second result. $\square$

Theorem 89 $\vdash \overset{\pi}{\neg}\tau \Rightarrow \left(\overset{\pi}{\neg}\sigma \Rightarrow \overset{\pi}{\neg}\varphi \right)$ *iff* $\vdash \left(\overset{\pi}{\neg}\tau \wedge \overset{\pi}{\neg}\sigma \right) \Rightarrow \overset{\pi}{\neg}\varphi.$ *(implication adjunction in the Boolean core L_π)*

Proof: Assuming the RHS, $\vdash \overset{\pi}{\neg}\overset{\pi}{\neg}\varphi \Rightarrow \overset{\pi}{\neg}\left(\overset{\pi}{\neg}\tau \wedge \overset{\pi}{\neg}\sigma \right)$ iff $\vdash \overset{\pi}{\neg}\overset{\pi}{\neg}\varphi \Rightarrow \left(\overset{\pi}{\neg}\overset{\pi}{\neg}\tau \vee \overset{\pi}{\neg}\overset{\pi}{\neg}\sigma \right)$ iff $\vdash \overset{\pi}{\neg}\overset{\pi}{\neg}\varphi \Rightarrow \left(\overset{\pi}{\neg}\tau \Rightarrow \overset{\pi}{\neg}\overset{\pi}{\neg}\sigma \right)$ iff $\vdash \overset{\pi}{\neg}\tau \Rightarrow \left(\overset{\pi}{\neg}\overset{\pi}{\neg}\varphi \Rightarrow \overset{\pi}{\neg}\overset{\pi}{\neg}\sigma \right)$ iff $\vdash \overset{\pi}{\neg}\tau \Rightarrow \left(\overset{\pi}{\neg}\sigma \Rightarrow \overset{\pi}{\neg}\varphi \right)$ which is the LHS. $\square$

Theorem 90 $\vdash \left(\overset{\pi}{\neg}\tau \vee \overset{\pi}{\neg}\varphi \right) \wedge \overset{\pi}{\neg}\sigma \Leftrightarrow \left(\overset{\pi}{\neg}\tau \wedge \overset{\pi}{\neg}\sigma \right) \vee \left(\overset{\pi}{\neg}\varphi \wedge \overset{\pi}{\neg}\sigma \right).$ *(distributivity of meet over join in the Boolean core L_π)*

Proof: Using Ore's theorem,

$$\vdash \left(\overset{\pi}{\neg}\tau \wedge \overset{\pi}{\neg}\sigma \right) \vee \left(\overset{\pi}{\neg}\varphi \wedge \overset{\pi}{\neg}\sigma \right) \Leftrightarrow \left(\overset{\pi}{\neg}\tau \vee \left(\overset{\pi}{\neg}\varphi \wedge \overset{\pi}{\neg}\sigma \right) \right) \wedge \left(\overset{\pi}{\neg}\sigma \vee \left(\overset{\pi}{\neg}\varphi \wedge \overset{\pi}{\neg}\sigma \right) \right).$$

And then by Ore's theorem again (which requires $\overset{\pi}{\neg}\sigma$ rather than σ),

$$\vdash \left[\left(\overset{\pi}{\neg}\tau \wedge \overset{\pi}{\neg}\sigma \right) \vee \left(\overset{\pi}{\neg}\varphi \wedge \overset{\pi}{\neg}\sigma \right) \right] \Leftrightarrow$$
$$(\overset{\pi}{\neg}\tau \vee \overset{\pi}{\neg}\varphi) \wedge \left(\overset{\pi}{\neg}\tau \vee \overset{\pi}{\neg}\sigma \right) \wedge \left(\overset{\pi}{\neg}\sigma \vee \overset{\pi}{\neg}\varphi \right) \wedge \left(\overset{\pi}{\neg}\sigma \vee \overset{\pi}{\neg}\sigma \right).$$

Then we have:

$$\vdash \left[\left(\overset{\pi}{\neg}\tau \wedge \overset{\pi}{\neg}\sigma \right) \vee \left(\overset{\pi}{\neg}\varphi \wedge \overset{\pi}{\neg}\sigma \right) \right] \Leftrightarrow \left((\overset{\pi}{\neg}\tau \vee \overset{\pi}{\neg}\varphi) \wedge \overset{\pi}{\neg}\sigma \right) \wedge \left[\left(\overset{\pi}{\neg}\tau \wedge \overset{\pi}{\neg}\varphi \right) \vee \overset{\pi}{\neg}\sigma \right]$$

but $\vdash \overset{\pi}{\neg}\sigma \Rightarrow \left[\left(\overset{\pi}{\neg}\tau \wedge \overset{\pi}{\neg}\varphi \right) \vee \overset{\pi}{\neg}\sigma \right]$ so that $\vdash \left[\left(\overset{\pi}{\neg}\tau \wedge \overset{\pi}{\neg}\varphi \right) \vee \overset{\pi}{\neg}\sigma \right] \wedge \overset{\pi}{\neg}\sigma \Leftrightarrow \overset{\pi}{\neg}\sigma$ and thus:

$$\vdash \left[\left(\overset{\pi}{\neg}\tau \wedge \overset{\pi}{\neg}\sigma \right) \vee \left(\overset{\pi}{\neg}\varphi \wedge \overset{\pi}{\neg}\sigma \right) \right] \Leftrightarrow \left((\overset{\pi}{\neg}\tau \vee \overset{\pi}{\neg}\varphi) \wedge \overset{\pi}{\neg}\sigma \right). \qquad \square$$

Theorem 91 $\vdash \overset{\pi}{\neg}\tau \Rightarrow \left(\overset{\pi}{\neg}\sigma \Rightarrow \left(\overset{\pi}{\neg}\tau \wedge \overset{\pi}{\neg}\sigma \right) \right).$ *(accumulation in the Boolean core L_π)*

Proof: In the adjunction theorem, replace $\overset{\pi}{\neg}\varphi$ by $\left(\overset{\pi}{\neg}\tau \wedge \overset{\pi}{\neg}\sigma \right) \Leftrightarrow \overset{\pi}{\neg}\left(\overset{\pi}{\neg}\overset{\pi}{\neg}\sigma \vee \overset{\pi}{\neg}\overset{\pi}{\neg}\tau \right)$ so that the "bottom" is a theorem and thus the top is a theorem. $\square$

Corollary 31 $\vdash \left[\overset{\pi}{\neg}\sigma \Rightarrow \left[\overset{\pi}{\neg}\sigma \wedge \left(\overset{\pi}{\neg}\sigma \Rightarrow \overset{\pi}{\neg}\varphi \right) \right] \right] \Leftrightarrow \left[\overset{\pi}{\neg}\sigma \Rightarrow \overset{\pi}{\neg}\varphi \right].$

Proof: $\vdash$ $\left[\overset{\pi}{\neg}\sigma \Rightarrow \left[\overset{\pi}{\neg}\sigma \wedge \left(\overset{\pi}{\neg}\sigma \Rightarrow \overset{\pi}{\neg}\varphi \right) \right] \right]$ $\Leftrightarrow$ $\left[\overset{\pi}{\neg}\sigma \Rightarrow \left(\overset{\pi}{\neg}\sigma \wedge \overset{\pi}{\neg}\varphi \right) \right]$ $\Leftrightarrow$
$\overset{\pi}{\neg}\overset{\pi}{\neg}\sigma \vee \left(\overset{\pi}{\neg}\sigma \wedge \overset{\pi}{\neg}\varphi \right)$ and $\vdash \overset{\pi}{\neg}\overset{\pi}{\neg}\sigma \vee \left(\overset{\pi}{\neg}\sigma \wedge \overset{\pi}{\neg}\varphi \right) \Leftrightarrow \left(\overset{\pi}{\neg}\overset{\pi}{\neg}\sigma \vee \overset{\pi}{\neg}\sigma \right) \wedge \left(\overset{\pi}{\neg}\overset{\pi}{\neg}\sigma \vee \overset{\pi}{\neg}\varphi \right) \Leftrightarrow$
$\left(\overset{\pi}{\neg}\overset{\pi}{\neg}\sigma \vee \overset{\pi}{\neg}\varphi \right) \Leftrightarrow \left[\overset{\pi}{\neg}\sigma \Rightarrow \overset{\pi}{\neg}\varphi \right].$ $\square$

It is an open question whether or not these axioms and rules of inference suffice to derive all partition tautologies. But we have given syntactic development to show how proofs work in partition logic as in any other logic.

Chapter 3

A dual structure on the algebra of partitions

3.1 Co-negation on partitions

The previous results about the 16 logical operations on partitions were restated as complementary-dual operations on equivalence relations. This produced no new operations on partitions (or partition relations or ditsets); it produced just a complementary viewpoint. Our topic now is the genuinely dual structure on the algebra of partitions $\Pi(U)$ that does define new operations on partitions. The structure on $\Pi(U)$ covered above based on negation and implication will be called the *primal structure* and the structure on $\Pi(U)$ based on co-negation and co-implication or difference is the *dual structure*.

In the analogy with Heyting algebras, this dual structure gives the partition version of co-negation and co-implication associated with co-Heyting algebras–so that the partition algebra is akin to a bi-Heyting or Heyting-Brouwer algebra except that it is more complicated by virtue of being non-distributive.

The previously developed partition negation $\neg\sigma = \sigma \Rightarrow \mathbf{0}_U$ might be associated with the "truth-value" of "Same as $\mathbf{0}_U$?" applied to σ. In other words, if $\sigma = \mathbf{0}_U$, then $\neg\sigma = \mathbf{1}_U$ since it is then true that σ is the "Same as $\mathbf{0}_U$", and otherwise when $\sigma \neq \mathbf{0}_U$, then $\neg\sigma = \mathbf{0}_U$. The dual co-negation is associated with the degree of "Difference from $\mathbf{1}_U$?" applied to σ. The indiscrete partition $\mathbf{0}_U$ has only one block U, while the discrete partition $\mathbf{1}_U$ has the maximum number of blocks $|U|$, so the co-negation "Difference from $\mathbf{1}_U$," denoted $\mathbf{1}_U - \sigma = -\sigma$ is a more varied measure than the negation $\sigma \Rightarrow \mathbf{0}_U = \neg\sigma$.

A partition is said to be *modular* if there is at most one block that is not a singleton and there is not just one singleton, so $\mathbf{1}_U$ and $\mathbf{0}_U$ are both modular partitions. The set-of-blocks definition of the *co-negation* $1_U - \sigma = -\sigma$ is the partition where all the singleton blocks of σ are collected together in one block and all the non-singleton blocks of σ are atomized into singletons. Thus the co-negated partitions are the same as the modular partitions.

In the context of co-Heyting algebras, Lawvere [68, p. 279] described the non-σ negation as the smallest element whose join with σ is the top element– just as the negation $\neg\sigma$ is the largest element whose meet with σ is the bottom element. Both the negation $\neg\sigma$ and co-negation $-\sigma$ satisfy that requirement even though the lattice of partitions $\Pi(U)$ (being non-distributive) is neither a Heyting nor a co-Heyting algebra.

For instance, if $\sigma = \{a, b, cd, ef\}$, then the co-negation is the modular partition $-\sigma = \{ab, c, d, e, f\}$ and the double co-negation is the modular partition $--\sigma = \{a, b, cdef\}$ with $--\sigma \precsim \sigma$. The co-negation of 1_U is $1_U - 1_U = 0_U$, the co-negation of 0_U is $1_U - 0_U = 1_U$, and for any $\sigma \neq 0_U, 1_U$, the co-negation $-\sigma$ is between the top and bottom of $\Pi(U)$ with the singletons in $-\sigma$ indicating the degree of σ's "difference from 1_U." That is, $1_U - 1_U = -1_U = 0_U$ indicates $\sigma = 1_U$ has no difference from 1_U (i.e., the "difference from 1_U" is maximally false) while $1_U - 0_U = 1_U$ indicates $\sigma = 0_U$ has complete difference from 1_U (i.e., the "difference from 1_U" is maximally true), and otherwise $1_U - \sigma = -\sigma$ is in between. Any σ with no singletons or a single singleton is also treated as completely different from 1_U so $-\sigma = 1_U$.

To see in more detail how the dualization works between negation and co-negation as well as to set up the relativized cases of π-negation and π-co-negation, consider the following characterizations of the two negations. For the negation $\neg\sigma$, i.e., "Same as 0_U", the definition is given in Table 3.1. The notation $(\sigma \vee 0_U)_U$ stands for the block in $\sigma \vee 0_U$ that contains U.

If $(\sigma \vee 0_U)_U \neq U$, then $(\neg\sigma)_U = (\sigma \Rightarrow 0_U)_U = U$
If $(\sigma \vee 0_U)_U = U$, then $(\neg\sigma)_U = (\sigma \Rightarrow 0_U)_U = 1_U \upharpoonright U$

Table 3.1: Definition of the negation $\sigma \Rightarrow 0_U = \neg\sigma$.

Then the dual version is given in Table 3.2 for co-negation, i.e., "Difference from 1_U".

If $(\sigma \wedge 1_U)_{\{u\}} \neq \{u\}$, then $(-\sigma)_{\{u\}} = (1_U - \sigma)_{\{u\}} = \{u\}$
If $(\sigma \wedge 1_U)_{\{u\}} = \{u\}$, then $(-\sigma)_{\{u\}} = (1_U - \sigma)_{\{u\}} = 0_U \upharpoonright \{u\}$

Table 3.2: Definition of the co-negation $1_U - \sigma = -\sigma$.

In Table 3.2, the blocks are singletons $\{u\} \in 1_U$, $(\sigma \wedge 1_U)_{\{u\}}$ stands for the block of $\sigma \wedge 1_U$ containing $\{u\}$ as a subset, and $0_U \upharpoonright \{u\}$ stands for the non-singleton block including u. Hence if $(\sigma \wedge \mathbf{1}_U)_{\{u\}} = \{u\}$, then $\{u\}$ was a singleton block of σ so $(\mathbf{1}_U - \sigma)_{\{u\}}$, the block of the co-negation containing u, is the non-singleton block represented by $\mathbf{0}_U \upharpoonright \{u\}$, and if $(\sigma \wedge \mathbf{1}_U)_{\{u\}} \neq \{u\}$, then $\{u\}$ was part of a bigger block in σ so it is turned into a singleton, i.e., the block of $(\mathbf{1}_U - \sigma)_{\{u\}}$ containing u is the singleton $\{u\}$.

The *co-regular* partitions are the co-negated elements of $\Pi(U)$, i.e., the modular partitions. We form an algebra $\mathcal{M}[\Pi(U)] = \{-\sigma : \sigma \in \Pi(U)\}$ of the modular partitions. Each modular partition ($\neq \mathbf{1}_U$) has the form $\{..., M, ...\}$ of one non-singleton block M, called the *modular block*, and otherwise two or more singleton blocks or no singleton blocks when $M = U$. The co-negation in the algebra just interchanges the modular block and the singletons, i.e., the modular block is atomized into singletons and the singletons are collected together to form the modular block of the co-negation. The top of the algebra is $\mathbf{1}_U$ and the bottom $\mathbf{0}_U$. The coatoms in $\mathcal{M}[\Pi(U)]$ are the partitions of the form $\{..., \{u, u'\}, ...\}$ and the atoms are their co-negations $\{\{u\}, U - \{u, u'\}, \{u'\}\}$. Given two co-regular partitions $\sigma = \{..., M, ...\}$ and $\tau = \{..., M', ...\}$, their join is the partition $\{..., M \cap M', ...\}$ which would be $\mathbf{1}_U$ if $M \cap M'$ was a singleton or empty, and that is the join $\sigma \vee \tau$ in $\Pi(U)$. If M and M' overlap, their meet in the algebra is the modular partition $\{..., M \cup M', ...\}$ which is the meet in $\Pi(U)$, but is otherwise not the meet $(..., M, M', ...)$ in $\Pi(U)$. In general, the modular meet $\{..., M \cup M', ...\}$ in the algebra is the double co-negation $--(\sigma \wedge \tau) = \{..., M \cup M', ...\}$.

Oystein Ore [79] showed that every partition π on U generates a Boolean algebra $\mathcal{B}(\pi)$, a complete subalgebra of $\wp(U)$, whose subsets are just the unions of blocks of π. In the case of $\pi = \mathbf{1}_U$, $\mathcal{B}(\mathbf{1}_U) = \wp(U)$. To relate $\mathcal{M}[\Pi(U)]$ to $\mathcal{B}(\mathbf{1}_U)$, we have to (as before) make a special treatment of the singletons. Since for $|U| = 2$, $\Pi(U)$ is already a Boolean algebra under the partition operations, we assume $|U| \geq 3$. The problem caused by singletons is that partitions of the form $\mu_u = \{\{u\}, U - \{u\}\}$ have a single non-singleton but are not the co-negation of any partitions due to the single singleton. Hence we need to make a number of changes to $\mathcal{M}[\Pi(U)]$ to create $\mathcal{M}[\Pi(U)]^*$ in order to relate it to $\mathcal{B}(\mathbf{1}_U) = \wp(U)$. We make the following changes:

- introduce all the $\mu_u = \{\{u\}, U - \{u\}\}$ partitions as the atoms in $\mathcal{M}[\Pi(U)]^*$;

- introduce a new layer of coatoms between previous coatoms $\{..., \{u, u'\}, ...\}$, the new coatoms being $|U|$ copies of the discrete partition but, for each $u \in U$, with one singleton $\{..., \{u\}^*, ...\}$ marked to act as the 'modular block' M which is then treated as the co-negation

$-\mu_u = -\{\{u\}, U - \{u\}\}$ so that $-\{..., \{u\}^*, ...\} = \{\{u\}, U - \{u\}\}$;

- in the refinement ordering $\precsim$, $\{..., M, ...\} \precsim \{..., \{u\}^*, ...\}$ iff $u \in M$; and

- when $M \cap M'$ is a singleton $\{u\}$, then the join is $\{..., M \cap M', ...\} = \{..., \{u\}^*, ...\}$, otherwise when $M \cap M' = \emptyset$, then $\{..., M \cap M', ...\}$ is the top $\mathbf{1}_U$.

The partition operations in $\mathcal{M}[\Pi(U)]^*$ correspond to the Boolean operations in $\mathcal{B}(\mathbf{1}_U) = \wp(U)$ as follows:

- Join: $(..., M, ...) \vee (...M', ...) = (..., M \cap M', ...)$ and if $M \cap M' = \{u\}$ then it is $(..., \{u\}^*, ...)$. If $M \cap M' = \emptyset$, then the join is $\mathbf{1}_U$.

- Meet: The modular-meet is $- -(..., M, ...) \wedge (...M', ...) = (..., M \cup M', ...)$.

- Co-negation: $-(..., M, ...) = (..., M^c, ...)$ where the singletons are the elements of M and $M^c = U - M$.

These changes form a new algebra $\mathcal{M}[\Pi(U)]^*$ which is a Boolean algebra anti-isomorphic to $\mathcal{B}(\mathbf{1}_U) = \wp(U)$ under the one-to-one correspondence: $\{..., M, ...\} \longleftrightarrow M \in \mathcal{B}(\mathbf{1}_U) = \wp(U)$:

$$\mathcal{M}[\Pi(U)]^{*op} \cong \mathcal{B}(\mathbf{1}_U) = \wp(U).$$

The Boolean algebra $\mathcal{M}[\Pi(U)]^*$ is not a sublattice of $\Pi(U)$ due to modular-meet operation (double co-negation of the ordinary meet), not to mention the added coatoms $(..., \{u\}^*, ...)$. Thus we cannot infer that Boolean tautologies in $\mathcal{M}[\Pi(U)]^*$ will be partition tautologies in $\Pi(U)$–as we could do for the Boolean cores in the primal structure (formed by $\vee$, $\wedge$, $\neg$, $\mathbf{0}_U$, and $\mathbf{1}_U$).

The Boolean algebra $\mathcal{B}(\mathbf{1}_U)$ is formed by taking all unions of the blocks of $\mathbf{1}_U$, i.e., all subsets $M \in \wp(U)$. The added coatoms $\{..., \{u\}^*, ...\}$ correspond to the singletons (the atoms of $\mathcal{B}(\mathbf{1}_U)$), the atomic partitions $\mu_u = \{\{u\}, U - \{u\}\}$ correspond to the coatoms of $\mathcal{B}(\mathbf{1}_U)$, and in general $\{..., M, ...\} \longleftrightarrow M \in \mathcal{B}(\mathbf{1}_U)$ to form the anti-isomorphism. The refinement ordering between modular partitions $(..., M, ...) \precsim (..., M', ...)$ iff $M' \subseteq M$ is why it is an order-reversing isomorphism.

For an example as to why $\mathcal{M}[\Pi(U)]$ is not a Boolean algebra without the added atoms and coatoms, consider the following example. All the partitions (written in shorthand) $\pi = \{a, d, bc\}$, $\sigma = \{b, d, ac\}$, and $\tau = \{ab, c, d\}$ are modular but $\pi \wedge \sigma = \{d, abc\}$ and $\pi \wedge \tau = \{d, abc\}$ which are not modular (since there on only one singleton). And the $\sigma \vee \tau = \mathbf{1}_U$ so:

$$\pi \wedge (\sigma \vee \tau) = \pi = \{a, d, bc\} \neq \{d, abc\} = (\pi \wedge \sigma) \vee (\pi \wedge \tau)$$

and thus distributivity fails. But when we add the coatoms $\{..., \{u\}^*, ...\}$ as the co-negations of the atomic partitions μ_u, then $\sigma \vee \tau = \{b, d, ac\} \vee \{ab, c, d\} = \{a^*, b, c, d\}$ (since $M \cap M' = \{a\}$). And then

$$\begin{aligned} \pi \wedge (\sigma \vee \tau) &= \{a, d, bc\} \wedge \{a^*, b, c, d\} = \{abc, d\} \\ &= \{d, abc\} \vee \{d, abc\} = (\pi \wedge \sigma) \vee (\pi \wedge \tau) \end{aligned}$$

so distributivity is not violated. The corresponding operations in $\wp(U)$ under the anti-isomorphism are:

$$\begin{aligned} \{b, c\} \cup (\{a, c\} \cap \{a, b\}) &= \{b, c\} \cup \{a\} = \{a, b, c\} \text{ and} \\ (\{b, c\} \cup \{a, c\}) \cap (\{b, c\} \cup \{a, b\}) &= \{a, b, c\} \cap \{a, b, c\} = \{a.b.c\} \end{aligned}$$

which are the elements in the modular block of $\{abc, d\}$.

3.2 Some tautologies for co-negation

We have extensively studied the tautologies for the negation $\sigma \Rightarrow \mathbf{0}_U = \neg\sigma$ and the π-relative negation or implication $\overset{\pi}{\neg}\sigma = \sigma \Rightarrow \pi$. Here only some of the new tautologies for co-negation will be mentioned.

Perhaps the most obvious difference is that the law of excluded middle holds for co-negation so it is a partition tautology:

$$\sigma \vee -\sigma = \mathbf{1}_U.$$

The proof is simply that every block of σ that is not a singleton will be atomized into singletons in $-\sigma$ so the join is all singletons, i.e., $\mathbf{1}_U$.

In the context of co-Heyting algebras ([68], [67]), Lawvere calls $--\sigma$ the *core* of σ and $\sigma \wedge -\sigma$ the *boundary* $\partial\sigma$ of σ and then notes that σ is reconstructed as the join of its core and boundary, and this holds for co-negation in partition algebras as well:

$$--\sigma \vee \partial\sigma = \sigma.$$

In the case of the example $\sigma = \{a, b, cd, ef\}$, we have $-\sigma = \{ab, c, d, e, f\}$ and $--\sigma = \{a, b, cdef\}$ so that the boundary is $\partial\sigma = \sigma \wedge -\sigma = \{ab, cd, ef\}$ and $\sigma = --\sigma \vee \partial\sigma$.

The proofs requires us to keep track of a few generic types of blocks for any partition σ:

- the Original Singletons of σ, i.e., $os\,(\sigma)$;

- the Original Non-Singletons of σ, i.e., $ons\,(\sigma)$;

- the Modular block obtained by joining the Original Singletons of σ, i.e., $mos\,(\sigma)$ (unless $|os\,(\sigma)| = 1$);

- the Singletons obtained by atomizing the Original Non-Singletons of σ or $sons\,(\sigma)$; and

- the Modular block formed by the union of the Original Non-Singletons of σ or $mons\,(\sigma)$.

Then we can characterize some of the partitions formed by co-negation as follows:

$$\sigma = \{os\,(\sigma)\,,ons\,(\sigma)\}$$
$$-\sigma = \{mos\,(\sigma)\,,sons\,(\sigma)\}$$

Complications arise when there is a single singleton, i.e., $|os\,(\sigma)| = 1$. Then $os\,(\sigma) = mos\,(\sigma)$ so that $-\sigma = \mathbf{1}_U$. Then the treatment of the double co-negation splits into two cases:

$$- - \sigma = \begin{cases} \{os\,(\sigma)\,,mons\,(\sigma)\} & \text{if } |os\,(\sigma)| \neq 1 \\ \mathbf{0}_U & \text{if } |os\,(\sigma)| = 1 \end{cases}.$$

Note that in either case, the triple co-negation is the same as the single co-negation:

$$- - -\sigma = -\sigma.$$

And in either case, the boundary of σ is:

$$\partial\sigma = \sigma \wedge -\sigma = \{mos\,(\sigma)\,,ons\,(\sigma)\}.$$

The proof the law of excluded middle for co-negation is then:

$$\sigma \vee -\sigma = \{os\,(\sigma)\,,ons\,(\sigma)\} \vee \{mos\,(\sigma)\,,sons\,(\sigma)\} = \{os\,(\sigma)\,,sons\,(\sigma)\} = \mathbf{1}_U.$$

Then the proof of Lawvere's core plus boundary result, $- - \sigma \vee \partial\sigma = \sigma$, consists in observing that:

$$- - \sigma \vee \partial\sigma = \{os\,(\sigma)\,,mons\,(\sigma)\} \vee \{mos\,(\sigma)\,,ons\,(\sigma)\} = \{os\,(\sigma)\,,ons\,(s)\} = \sigma$$

or if $mos(\sigma) = os(\sigma)$, i.e., $|os(\sigma)| = 1$, then:

$$--\sigma \vee \partial\sigma = \mathbf{0}_U \vee \{mos(\sigma), ons(\sigma)\} = \{os(\sigma), ons(s)\} = \sigma.$$

Furthermore, the boundary $\partial(--\sigma)$ of $--\sigma$ is the same as the boundary $\partial(-\sigma)$ of $-\sigma$:

$$\partial(--\sigma) = --\sigma \wedge ---\sigma = --\sigma \wedge -\sigma = \partial(-\sigma) = \{mos(\sigma), mons(\sigma)\}.$$

Thus $\partial(-\sigma)$ is refined by $\partial\sigma$ so $\partial(-\sigma) \Rightarrow \partial\sigma$ is a partition tautology. For negation, $\sigma \precsim \neg\neg\sigma$ so we might expect $--\sigma \precsim \sigma$ to hold in the dual structure which means that $--\sigma \Rightarrow \sigma$ is a partition tautology. The proof is simply the observation that $--\sigma = \{os(\sigma), mons(\sigma)\} \precsim \{os(\sigma), ons(\sigma)\} = \sigma$ or $--\sigma = \mathbf{0}_U \precsim \sigma$. Then $\sigma = --\sigma$ if and only if $mons(\sigma) = ons(\sigma)$ and $mos(\sigma) \neq os(\sigma)$, so that σ has only one non-singleton block and not just one singleton block, i.e., when σ is modular.

These calculations for co-negation might be simplified if we translate them into ditsets. The generic structure of any partition σ is $\{os(\sigma), ons(\sigma)\}$. The ditset $dit(\sigma)$ is made up of three disjoint sets of distinctions which are numbered as follows:

1. $\cup_{\substack{C,C' \in os(\sigma) \\ C \neq C'}} C \times C'$ which might be symbolized as $dit(os(\sigma), os(\sigma))$;

2. $\cup_{\substack{C \in os(\sigma) \\ C' \in ons(\sigma)}} C \times C' \cup \cup_{\substack{C \in os(\sigma) \\ C' \in ons(\sigma)}} C' \times C$ which might be symbolized as $dit(os(\sigma), ons(\sigma))$; and

3. $\cup_{\substack{C,C' \in ons(\sigma) \\ C \neq C'}} C \times C'$ which might be symbolized as $dit(ons(\sigma), ons(\sigma))$.

These three sets of distinctions are disjoint and their union is $dit(\sigma)$ which might be symbolized as $\sigma = 1 + 2 + 3$. The only remaining distinctions are those that would be added when all the non-singletons blocks of σ are atomized into singletons and they form the fourth set of distinctions disjoint from the other three sets:

4. $\cup_{C \in ons(\sigma)} C \times C - \Delta$.

Together these four disjoint set of distinction exhaust all possible distinction so their union is 1_U. When σ is co-negated to form $-\sigma = \{mos(\sigma), sons(\sigma)\}$, then the set 1 is eliminated and the set 4 is added on while the sets 2 and 3 remain so the ditset dit $(-\sigma)$ of $-\sigma$ is the union of the sets 2, 3, and 4 which is symbolized as: $-\sigma = 2+3+4$. The double co-negation $--\sigma$ of σ has the form $--\sigma = \{os\,(\sigma)\,,mons\,(\sigma)\}$, so set 1 is added back and the sets 3 and 4 are eliminated which is symbolized as $--\sigma = 1+2$. All the partitions that can be formed from σ by the lattice operations and co-negation will have two possible sets of blocks $os\,(\sigma)$ or $mos\,(\sigma)$ for the singletons and three possible sets of blocks $ons\,(\sigma)$, $sons\,(\sigma)$, and $mons\,(\sigma)$ so there are six possible partitions that can be thus formed, and they are illustrated in Figure 3.1.

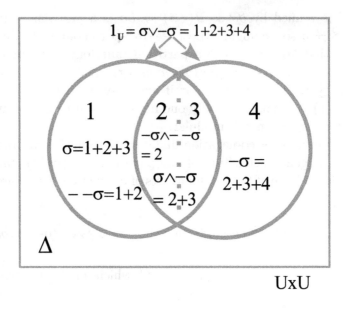

Figure 3.1: The Venn diagram for the six partitions formed from σ by $\vee$, $\wedge$, and $-$.

Then the core + boundary result is simply that the union of the core ditsets $1+2$ and the boundary ditsets $2+3$ is the ditset $1+2+3$ for σ.
Table 3.3 lists the six partitions.

$\sigma \vee -\sigma = 1_U$
$\sigma = 1+2+3$
$-\sigma = 2+3+4$
$--\sigma = 1+2$
$\sigma \wedge -\sigma = 2+3$
$-\sigma \wedge --\sigma = 2$

Table 3.3: The six partitions formed from σ by $\vee$, $\wedge$, and $-$.

Recall that a partition σ is modular if it has at most one non-singleton and not just one singleton. All co-negated partitions are modular and a partition $\sigma \neq \mathbf{0}_U$ is modular if and only if it equals its double co-negation, $\sigma = --\sigma$. By comparing the formulas in Table 3.3, we see that a partition is modular if and only if the set of distinctions #3 is empty and #1 is not empty.

Since the refinement of partitions is just inclusion between ditsets, we immediately have the refinements: $--\sigma \precsim \sigma, \sigma \wedge -\sigma \precsim \sigma, -\sigma$ (which also follows from the meet being the greatest lower bound), and $-\sigma \wedge --\sigma$ is refined by the other five partitions. Since the ditset for the join of two partitions is the union of the ditsets, we can represent the joins by just 'adding the numbers.' For instance, that gives immediately; $-\sigma \vee --\sigma = 1+2+3+4 = \mathbf{1}_U = \sigma \vee -\sigma$ and $--\sigma \vee (\sigma \wedge -\sigma) = 1+2+3 = \sigma$. The ditset of the meet of two partitions is the interior of the intersection of the two ditsets–which is the largest ditset contained in that intersection. The intersection of two ditsets of partitions is a set of distinctions but not necessarily the ditset of a partition. But in the case of the six partitions of Table 3.3, all the intersections, which will always be just a sum of the numbered sets, are the ditsets of one of the six functions so they are closed under meets. The co-negation of the double co-negation $--\sigma = \{os(\sigma), mons(\sigma)\}$ is $\{mos(\sigma), sons(\sigma)\}$ so in terms of the numbered sets, the triple co-negation eliminates set 1 and adds the sets 3 and 4 so it is the same as the single co-negation $-\sigma$. The co-negation $-(\sigma \wedge -\sigma)$ of $\sigma \wedge -\sigma = \{mos(\sigma), ons(\sigma)\}$ is $\mathbf{1}_U$ as will be proven just below using the strong DeMorgan law $-(\sigma \wedge \pi) = -\sigma \vee -\pi$. And by the same law, the co-negation $-(-\sigma \wedge --\sigma) = --\sigma \vee ---\sigma = --\sigma \vee -\sigma = \mathbf{1}_U$. Thus the six partitions are also closed under co-negation.

With two partitions, the first thing to check is DeMorgan's laws for co-negation. For negation, the 'weak' DeMorgan law $\neg(\sigma \vee \pi) = \neg\sigma \wedge \neg\pi$ holds but the 'strong' law $\neg(\sigma \wedge \pi) = \neg\sigma \vee \neg\pi$ fails. But for co-negation, the weak law fails as shown by the counterexample of $\sigma = \{a, b, cd, ef\}$ and $\pi = \{a, bc, def\}$. Then $\sigma \vee \pi = \{a, b, c, d, ef\}$ and $-(\pi \vee \sigma) = \{abcd, e, f\}$. On the RHS of the weak law, $-\sigma = \{ab, c, d, e, f\}$ and $-\pi = \mathbf{1}_U$ so:

$$-\sigma \wedge -\pi = -\sigma = \{ab, c, d, e, f\} \neq \{abcd, e, f\} = -(\pi \vee \sigma).$$

To analyze the strong DeMorgan law, we need to develop some more of the standard forms for co-negation with two partitions.

$$\sigma \wedge \pi = \{os(\sigma) \cap os(\pi), ons(\sigma \wedge \pi)\}$$
$$-(\sigma \wedge \pi) = \{m[os(\sigma) \cap os(\pi)], sons(\sigma \wedge \pi)\}$$
$$-\sigma \vee -\pi = \{mos(\sigma) \cap mos(\pi), sons(\sigma) \cup sons(\pi)\}.$$

Then to prove the strong law, we first note that $m \left(os \left(\sigma \right) \cap os \left(\pi \right) \right) = mos \left(\sigma \right) \cap mos \left(\pi \right)$ since the only singletons in $\sigma \wedge \pi$ are the ones common to both partitions. Then to show $sons \left(\sigma \wedge \pi \right) = sons \left(\sigma \right) \cup sons \left(\pi \right)$, we note that the meet operation only expands any block so non-singleton blocks remain non-singleton and thus the singletons $sons \left(\sigma \wedge \pi \right)$ made from the non-singleton blocks of $\sigma \vee \pi$ will be the same as $sons \left(\sigma \right) \cup sons \left(\pi \right)$ which completes the proof of the strong DeMorgan law for co-negation:

$$- \left(\sigma \wedge \pi \right) = - \sigma \vee - \pi.$$

Thus co-negation reverses the situation for negation with respect to the two DeMorgan laws.

In the co-Heyting context, Lawvere notes that the Leibniz law (e.g., for derivatives of products) holds which in our case would be: $\partial \left(\sigma \wedge \pi \right) = \left(\partial \sigma \wedge \pi \right) \vee \left(\sigma \wedge \partial \pi \right)$. The simple proof of Leibniz's law in a co-Heyting algebra uses distributivity so one might expect it to be false for co-negation, and indeed a counterexample is $\pi = \{a, b, c, def\}$ so $-\pi = \{abc, d, e, f\}$ and for $\sigma = \{a, b, d, cef\}$, then $-\sigma = \{abd, c, e, f\}$. Hence $-\sigma \vee -\pi = \{ab, c, d, e, f\}$ and $\sigma \wedge \pi = \{a, b, cdef\}$. Then the LHS is: $\partial \left(\pi \wedge \sigma \right) = \{ab, cdef\}$. And $\partial \sigma = \sigma \wedge -\sigma = \{abd, cef\}$ so $\partial \sigma \wedge \pi = \{U\} = \mathbf{0}_U$ and $\partial \pi = \{abc, def\}$ so $\partial \pi \wedge \sigma = \{U\} = \mathbf{0}_U$ so the RHS is $\{U\} = \mathbf{0}_U$ and thus the Leibniz law fails for co-negation.

3.3   Difference operation on partitions: Relativizing co-negation to π

In the primal structure, the negation $\neg \sigma = \sigma \Rightarrow \mathbf{0}_U$ was the partition representing the extent to which "σ is the same as $\mathbf{0}_U$", and the implication $\sigma \Rightarrow \pi$, obtained by relativizing the negation to π, is the partition representing the extent to which "σ is refined by π" indicated in the Table 3.4. When $\pi = \mathbf{0}_U$, then "σ is refined by π" is the same as "σ is the same as $\mathbf{0}_U$" since the only partition that $\mathbf{0}_U$ refines is itself (refinement being a partial order). The notation $\left(\sigma \vee \pi \right)_B$ represents any block of the join $\sigma \vee \pi$ containing part or all of B.

#1.If $\left(\sigma \vee \pi \right)_B \neq B$, then $\left(\overset{\pi}{\neg} \sigma \right)_B = \left(\sigma \Rightarrow \pi \right)_B = B$
#2.If $\left(\sigma \vee \pi \right)_B = B$, then $\left(\overset{\pi}{\neg} \sigma \right)_B = \left(\sigma \Rightarrow \pi \right)_B = \mathbf{1}_U \upharpoonright B$

Table 3.4: Definition of the implication $\sigma \Rightarrow \pi = \overset{\pi}{\neg} \sigma$ as negation relativized to π.

In the dual structure, the co-negation $-\sigma = \mathbf{1}_U - \sigma$ is the partition representing the extent to which "σ is different from $\mathbf{1}_U$", and the co-implication or difference operation $\overset{\pi}{-}\sigma = \pi - \sigma$, obtained by relativizing the co-negation to π, is the partition representing the extent to which "σ is not refined by π" indicated in Table 3.5 (where the notation $(\sigma \wedge \pi)_B$ represents the block of the meet $\sigma \wedge \pi$ containing the block $B \in \pi$).

#1.If $(\sigma \wedge \pi)_B \neq B$, then $\left(\overset{\pi}{-}\sigma\right)_B = (\pi - \sigma)_B = B$
#2.If $(\sigma \wedge \pi)_B = B$, then $\left(\overset{\pi}{-}\sigma\right)_B = (\pi - \sigma)_B = \mathbf{0}_U \restriction B$

Table 3.5: Definition of the difference operation $\pi - \sigma = \overset{\pi}{-}\sigma$ as co-negation relativized to π.

The dual structure defined by π lives in the lower segment $[\mathbf{0}_U, \pi]$ of partitions refined by π. If we treat the blocks $B \in \pi$ as points in a universe set, then the lower segment is isomorphic to $\Pi(\pi) \cong [\mathbf{0}_U, \pi]$ [49, p. 192] where π as a set of blocks plays the role of the universe set of points. Hence the difference $\pi - \sigma$ will be essentially like $\mathbf{1}_\pi - \sigma$ with the blocks $B \in \pi$ playing the role of the singletons $\{B\} \in \mathbf{1}_\pi$. The two rules for determining $\pi - \sigma$ can be restated as follows.

1. If $(\sigma \wedge \pi)_B \neq B$, then the block of $\sigma \wedge \pi$ containing B is larger than B (analogous to a block of σ being a non-singleton in the case of $\mathbf{1}_U - \sigma$), so that block of $\sigma \wedge \pi$ is 'atomized' into the blocks B of π contained in the 'non-singleton' block of $\sigma \wedge \pi$ (analogous to a non-singleton block of σ being atomized into singletons in $\mathbf{1}_U - \sigma$ except that the blocks of π are playing the role of the singletons). That is, if $(\sigma \wedge \pi)_B \neq B$, then the block of $\sigma \wedge \pi$ containing B is larger than B so B stays the same block in $\pi - \sigma$.

2. If $(\sigma \wedge \pi)_B = B$, then B is an exact union of blocks $C \in \sigma$ (and if $\sigma \precsim \pi$, then $B = C$), and then the B-slot $\left(\overset{\pi}{-}\sigma\right)_B = (\pi - \sigma)_B$ in $\overset{\pi}{-}\sigma = \pi - \sigma$ is part of the union of such blocks from π, to form the 'π-modular' block of $\pi - \sigma$, analogous to the modular block when $\pi = \mathbf{1}_U$. That is, if $(\sigma \wedge \pi)_B = B$, then B is an exact union of blocks $C \in \sigma$, then B joins into the π-modular block of $\pi - \sigma$.

A partition σ is said to be π -*modular* if it is refined by π, has at most one block consisting of a union of blocks of π, and does not just have one singleton block B, where the π -*modular block* (when $\sigma \neq \pi$) is the block consisting of the union of two or more blocks of π. Let $\mathcal{M}\left[\Pi\left(\pi\right)\right]$ be the π-modular partitions of $[\mathbf{0}_U, \pi]$ with the refinement ordering. As in the case of $\mathbf{1}_U - \sigma$, the partitions that are not differences $\pi - \sigma$ are the ones of the form $\mu_B = \left\{B, \cup_{B' \in \pi; B \neq B'} B'\right\}$. And, as before, if we start with the set $\mathcal{M}\left[\Pi\left(\pi\right)\right]$ of all π-modular partitions and then add the new atoms μ_B and the artificial coatoms $\{..., B^*, ...\}$ and make the other corresponding changes so that $\{..., B^*, ...\} = \pi - \mu_B = \overset{\pi}{-}\mu_B$, then we obtain a Boolean algebra anti-isomorphic to $\mathcal{B}\left(\pi\right)$.

The *difference, co-implication,* or π-*co-negation* operation $\pi - \sigma = \overset{\pi}{-}\sigma$ creates a π-modular partition in the lower segment $[\mathbf{0}_U, \pi]$ of $\Pi\left(U\right)$. If σ is not in that lower segment, then the π-co-negation of $\sigma \wedge \pi$ is the same as the π-co-negation of σ since only $\sigma \wedge \pi$ occurs in rules #1 and #2. In other words, $\pi - \sigma = \pi - \left(\sigma \wedge \pi\right)$. For instance, if $\pi = \{a, b, cd, ef, gh\}$ and $\sigma = \{abgh, cd, ef\}$, then $\pi - \sigma = \overset{\pi}{-}\sigma = \{a, b, cdef, gh\}$ since $\{abgh\}$ is a union of blocks of π it is 'atomized' into the respective blocks of π as $\{a, b, gh\}$ and since $\{cd, ef\}$ are two singletons as blocks of π, they are combined into the π-modular block $cdef$ of $\pi - \sigma$. And $\sigma \wedge \pi = \{abgh, cd, ef\}$ so that $\pi - \left(\sigma \wedge \pi\right) = \overset{\pi}{-}\left(\sigma \wedge \pi\right) = \{a, b, cdef, gh\} = \pi - \sigma$.

For any $\sigma \in [\mathbf{0}_U, \pi]$, the difference operation is like co-negation in the form $\mathbf{1}_\pi - \sigma$ (thinking of the $B \in \pi$ as if they were singletons) so the results for $\mathbf{1}_U - \sigma = -\sigma$ can be applied *mutatis mutandis*. Thus for $\sigma \in [\mathbf{0}_U, \pi]$, the law of excluded middle in $[\mathbf{0}_U, \pi]$ takes the form of the partition tautology: $\sigma \vee \overset{\pi}{-}\sigma = \pi$.

The ditset analysis, developed above for co-negation, extends to π-co-negation. The π-co-negated partition is in the lower segment $[\mathbf{0}_U, \pi]$ so its ditset will always be contained in dit $\left(\pi\right)$. Any $\sigma \in [\mathbf{0}_U, \pi]$ will have the form of some original π-singletons $o\pi s \left(\sigma\right)$ in the form $B \in \pi$ and some original π-non-singletons $o\pi ns \left(\sigma\right)$ in the form $B \cup B' \cup ...$ so it has the form:

$$\sigma = \{o\pi s \left(\sigma\right), o\pi ns \left(\sigma\right)\}.$$

Then in Table 3.6 we apply the #1 and #2 rules to get:

$$\pi - \sigma = \overset{\pi}{-}\sigma = \{mo\pi s \left(\sigma\right), so\pi ns \left(\sigma\right)\}.$$

$1 = \text{dit}\left(o\pi s\left(\sigma\right), o\pi s\left(\sigma\right)\right) = \cup_{\substack{B, B' \in o\pi s\left(\sigma\right) \\ B \neq B'}} B \times B'$
$2 = \text{dit}\left(o\pi s\left(\sigma\right), o\pi ns\left(\sigma\right)\right) = \cup_{\substack{B \in o\pi s\left(\sigma\right) \\ B' \subseteq C' \in o\pi ns\left(\sigma\right)}} B \times B' \cup ...$
$3 = \text{dit}\left(o\pi ns\left(\sigma\right), o\pi ns\left(\sigma\right)\right) = \cup_{\substack{B \subseteq C \in o\pi ns\left(\sigma\right) \\ B' \subseteq C' \in o\pi ns\left(\sigma\right), C \neq C', B \neq B'}} B \times B'$
$4 = \cup_{\substack{B, B' \subseteq C \in o\pi ns\left(\sigma\right) \\ B \neq B'}} B \times B'$

Table 3.6: Ditset analysis for π-co-negation.

This is illustrated in Figure 3.2.

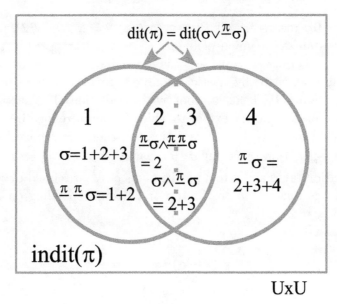

Figure 3.2: Ditset analysis π-co-negations of a partition σ.

Hence we have the Table 3.7 ditset analysis of the π-negated elements that can be formed from a single $\sigma \in [\mathbf{0}_U, \pi]$.

$\sigma \vee \overset{\pi}{-}\sigma = \sigma \vee (\pi - \sigma) = 1 + 2 + 3 + 4 = \pi$
$\sigma = 1 + 2 + 3$
$\overset{\pi}{-}\sigma = \pi - \sigma = 2 + 3 + 4$
$\overset{\pi}{-}\overset{\pi}{-}\sigma = \pi - (\pi - \sigma) = 1 + 2$
$\sigma \wedge \overset{\pi}{-}\sigma = \sigma \wedge (\pi = \sigma) = 2 + 3$
$\overset{\pi}{-}\sigma \wedge \overset{\pi}{-}\overset{\pi}{-}\sigma = 2$

Table 3.7: Formulas generated by π-negation of σ where $\sigma \precsim \pi$.

With π-negation, the partition $\sigma \Rightarrow \pi$ is the partition indicating the extent to which "σ is refined by π" so when it is totally true, we have refinement:

$$\overset{\pi}{\neg}\sigma = \sigma \Rightarrow \pi = \mathbf{1}_U \text{ iff } \sigma \precsim \pi.$$

For π-co-negation, we have the dual result (which also holds in co-Heyting algebras [87]):

$$\overset{\pi}{-}\sigma = \pi - \sigma = \mathbf{0}_U \text{ iff } \pi \precsim \sigma$$

since $\pi - \sigma = \mathbf{0}_U$ means for all $B \in \pi$, $(\sigma \wedge \pi)_B = B$ (the #2 rule above) so every $B \in \pi$ is an exact union of blocks $C \in \sigma$, i.e., $\pi \precsim \sigma$. In terms of the interpretation, when $\pi - \sigma = \mathbf{0}_U$ so "π is not refined by σ" is totally false, i.e., is a 'contradiction', then (by classical double negation), $\pi \precsim \sigma$. In that sense, showing that $\pi - \sigma = \mathbf{0}_U$ is like a proof by contradiction. The partition $\pi - \sigma$ is the partition representing the extent to which "π is not refined by σ" and when "π is not refined by σ" is the contradiction $\mathbf{0}_U$, then π is refined by σ, $\pi \precsim \sigma$.

Since the π-negation takes any σ to a partition $\sigma \Rightarrow \pi = \overset{\pi}{\neg}\sigma$ refining π and the π-co-negation takes any σ to a partition $\pi - \sigma = \overset{\pi}{-}\sigma$ refined by π, the mixed double negations take any σ to the top or bottom depending on the order of the negations:

$$\mathbf{1}_U = (\pi - \sigma) \Rightarrow \pi = \overset{\pi}{\neg}\overset{\pi}{-}\sigma$$

$$\mathbf{0}_U = \pi - (\sigma \Rightarrow \pi) = \overset{\pi}{-}\overset{\pi}{\neg}\sigma.$$

Another partition equation that gives some of the meaning of difference-from-π as subtraction is:

$$\pi \vee \sigma = (\pi - \sigma) \vee \sigma.$$

That is, to form $\pi \vee \sigma$, you could take σ away from π and then join σ back to get the same result. The partition join $\pi \vee \sigma$ is the partition formed by all the non-empty intersections $B \cap C$ for $B \in \pi$ and $C \in \sigma$. The difference $\pi - \sigma$ is formed using two rules. In the #2 rule, if $(\pi \wedge \sigma)_B = B$, then B is an exact union of some blocks $C \in \sigma$ so B would be joined into the π-modular block of $\pi - \sigma$, but then all the intersections of any $C \in \sigma$ with that block would give the same $B \cap C$ as in $\pi \vee \sigma$. In the #1 rule, $(\pi \wedge \sigma)_B \neq B$, then B is part of a larger block in the meet $\pi \wedge \sigma$, and thus B remains as the block B in $\pi - \sigma$ so the intersection $B \cap C$ is the same in both $\pi \vee \sigma$ and $(\pi - \sigma) \vee \sigma$.

A formula in lattice theory (with $\mathbf{0}_U$ and $\mathbf{1}_U$) is dualized by keeping the atomic variables the same, interchanging join and meet, and interchanging $\mathbf{0}_U$ and $\mathbf{1}_U$. Identities dualize into identities, e.g., $\pi \vee (\pi \wedge \sigma) = \pi$ dualizes to $\pi \wedge (\pi \vee \sigma) = \pi$. Dualization of formulas extends to the implication and co-implication by dualizing $\sigma \Rightarrow \pi$ to $\pi - \sigma$ and vice-versa, but the relationship between a formula involving the implications and its dual is more subtle.

This dualization should be distinguished from the complementary duality between partitions, represented by partition relations on $U \times U$, and equivalence relations on $U \times U$. The complementary dual of a statement about partitions is an equivalent statement about equivalence relations. The lattice-theoretic dualization and the extension to implication and co-implication is not that sort of rather trivial complementary duality. The dualization of a partition statement considered now is a different statement about partitions, not just an equivalent restatement in terms of equivalence relations.

A formula that equals $\mathbf{0}_U$ for any partitions substituted for the atomic variables and for any U ($|U| \geq 2$) is a *partition contradiction*, just as a formula that equals $\mathbf{1}_U$ for any substitution is a partition tautology.

The proof that any partition tautology was also a subset (or truth-table) tautology used the isomorphism $\wp(1) \cong \Pi(2)$, and the same proof strategy shows that any partition contradiction is also a subset or truth-table contradiction (i.e., the negation of a tautology). The subset operation that corresponds to the partition difference or co-implication operation $\pi - \sigma$ is the subset difference $S - T = S \cap T^c$ or in terms of propositional variables, $\pi \wedge \neg\sigma$. The set $U = 2 = \{0, 1\}$ allows only two partitions in $\Pi(2)$, namely $\mathbf{0}_U = \{\{0, 1\}\}$ and $\mathbf{1}_U = \{\{0\}, \{1\}\}$. In the isomorphism $\wp(1) \cong \Pi(2)$ between the two-element Boolean algebra and the partition algebra on the two-element set 2, the partition difference operation on $\mathbf{0}_U$ and $\mathbf{1}_U$ are isomorphic to the Boolean operations on zero and one of $\wp(1)$ as indicated in the following truth-table Table 3.8 (the the zeros and ones are interpreted as $\mathbf{0}_U$ and $\mathbf{1}_U$, or as $0, 1 \in U$ as the case may be).

π	σ	$\pi - \sigma$	$\pi \wedge \neg\sigma$
1	1	0	0
1	0	1	1
0	1	0	0
0	0	0	0

Table 3.8: Partition operation $\pi - \sigma$ in $\Pi(2)$ same as Boolean operation $\pi \wedge \neg\sigma$ in $\wp(1)$.

Since the other partition operations and Boolean operations are isomorphic in $\wp(1) \cong \Pi(2)$, any partition contradiction has to hold for $U = 2$, and thus the corresponding Boolean formula is also a contradiction so all partition contradictions are also Boolean contradictions.

The dual to the partition tautology of modus ponens $[\sigma \wedge (\sigma \Rightarrow \pi)] \Rightarrow \pi$ is the formula: $\pi - [\sigma \vee (\pi - \sigma)]$. But $\sigma \vee (\pi - \sigma) = \sigma \vee \pi$ and $\pi \precsim \sigma \vee \pi$ for any

π and σ, so we have:

$$\overset{\pi}{\neg}\left(\sigma \wedge \overset{\pi}{\neg}\sigma\right) = [\sigma \wedge (\sigma \Rightarrow \pi)] \Rightarrow \pi = \mathbf{1}_U$$

$$\overset{\pi}{-}\left(\sigma \vee \overset{\pi}{-}\sigma\right) = \pi - [\sigma \vee (\pi - \sigma)] = \mathbf{0}_U$$

which are the π-relativizations of:

$$\neg(\sigma \wedge \neg\sigma) = (\sigma \wedge (\sigma \Rightarrow \mathbf{0}_U)) \Rightarrow \mathbf{0}_U = \mathbf{1}_U$$

$$-(\sigma \vee -\sigma) = \mathbf{1}_U - [\sigma \vee (\mathbf{1}_U - \sigma)] = \mathbf{0}_U.$$

Modus ponens is a case where the dual of a partition tautology is a partition contradiction that is a different statement about partitions. Note that if a partition tautology has its main connective as an implication, then there is a trivial restatement as a contradiction using co-implication since both are equivalent to a certain refinement relation. Thus the modus ponens formula $[\sigma \wedge (\sigma \Rightarrow \pi)] \Rightarrow \pi = \mathbf{1}_U$ is equivalent to $[\sigma \wedge (\sigma \Rightarrow \pi)] \precsim \pi$ which in turn is equivalent to $[\sigma \wedge (\sigma \Rightarrow \pi)] - \pi = \mathbf{0}_U$, but that trivial restatement is not the dualization contradiction: $\pi - [\sigma \vee (\pi - \sigma)] = \mathbf{0}_U$.

There are many promising cases where identities dualize to identities. For instance, in the primal structure, the 'weak' DeMorgan law $\neg(\sigma \vee \tau) = \neg\sigma \wedge \neg\tau$ holds, and its dual, the 'strong' DeMorgan law, holds in the dual structure:

$$\neg(\sigma \vee \tau) = \neg\sigma \wedge \neg\tau$$

$$-(\sigma \wedge \tau) = -\sigma \vee -\tau.$$

If Φ is a formula in the primal structure ($\vee$, $\wedge$, $\Rightarrow$, $\mathbf{0}_U$, and $\mathbf{1}_U$), then let $\varphi(\Phi)$ be the dual formula in the dual structure ($\vee$, $\wedge$, $-$, $\mathbf{0}_U$, and $\mathbf{1}_U$). It is tempting to conjecture that the lattice theoretic dualization extends to implication and co-implication, i.e., to conjecture that: Φ is a partition tautology iff $\varphi(\Phi)$ is a partition contradiction. However, that is not true. There at least two special aspects of implication in the primal structure that are not mirrored in the dual structure so the implication, if Φ is a partition tautology, then $\varphi(\Phi)$ is a partition contradiction, does not hold.

One special aspect is the way that partial refinement is exemplified in the two structures. In the primal structure, and implication $\sigma \Rightarrow \pi$ indicated partial refinement when $B \subseteq C$ and then B is atomized into singletons in $\sigma \Rightarrow \pi$. In the partition tautology $\overset{\pi}{\neg}\sigma \vee \overset{\pi}{\neg}\overset{\pi}{\neg}\sigma$ (the weak law of excluded middle), there is a set of singletons in $\overset{\pi}{\neg}\sigma$ and a complementary set of singletons in $\overset{\pi}{\neg}\overset{\pi}{\neg}\sigma$ so the join is all singletons, i.e., $\mathbf{1}_U$. But in the dual structure, partial refinement in

$\pi - \sigma$ is indicated by $B = \cup C$ being added to the modular block of $\pi - \sigma$. In the dual formula, $\varphi\left(\overset{\pi}{\neg}\sigma \vee \overset{\pi}{\neg}\overset{\pi}{\neg}\sigma\right) = \overset{\pi}{-}\sigma \wedge \overset{\pi}{-}\overset{\pi}{-}\sigma$, the modular blocks will be complementary but their meet does not give $\mathbf{0}_U$. Taking the simple case of $\pi = \mathbf{1}_U$, and $\sigma = \{a, b, cd\}$, $\mathbf{1}_U - \sigma = -\sigma = \{ab, c, d\}$, $--\sigma = \{a, b, cd\}$, and $-\sigma \wedge --\sigma = \{ab, cd\} \neq \mathbf{0}_U$.

Another special aspect of the primal structure (due to the Common Dits Theorem), is the 'extremism' of negation, i.e., $\neg\sigma = \sigma \Rightarrow \mathbf{0}_U$ is either $\mathbf{1}_U$ (when $\sigma = \mathbf{0}_U$) or $\mathbf{0}_U$ otherwise. Consider the partition tautology, $\neg\sigma \Rightarrow (\sigma \Rightarrow \pi)$, which is the classical "reductio" that a contradiction implies anything. The reductio is a partition tautology since if $\sigma \neq \mathbf{0}_U$, then $\neg\sigma = \sigma \Rightarrow \mathbf{0}_U = \mathbf{0}_U$ so the whole expression is $\mathbf{0}_U \Rightarrow (\sigma \Rightarrow \pi) = \mathbf{1}_U$, and if $\sigma = \mathbf{0}_U$, then $\sigma \Rightarrow \pi = \mathbf{1}_U$ so the whole expression is also $\mathbf{1}_U$. The dual formula is, $(\pi - \sigma) - (\mathbf{1}_U - \sigma)$, so consider the case where $\pi = \{abe, cd\}$ and $\sigma = \{a, b, c, de\}$ so that $\sigma \wedge \pi = \mathbf{0}_U$ and thus $\pi - \sigma = \pi$ while $\mathbf{1}_U - \sigma = \{abc, d, e\}$. Then $(\pi - \sigma) - (\mathbf{1}_U - \sigma) = \pi - (\mathbf{1}_U - \sigma) = \pi \neq \mathbf{0}_U$ since $\pi \wedge (\mathbf{1}_U - \sigma) = \mathbf{0}_U$.

The opposite conjecture, if a formula Φ in the dual structure is a partition contradiction, then the formula $\varphi^{-1}(\Phi)$ in the primal structure is a partition tautology, is still open. The partition algebra with implication and coimplication does not exhibit all the nice symmetries of the H-B algebras (algebras that are both Heyting and Co-Heyting algebras) ([87]; [114]). The partition algebra is not distributive but it is defined in terms of partitions on arbitrary sets with no topology or ordering relations. Clearly much more work needs to be done to better understand the similarities and differences between partition algebras and H-B algebras.

Some of the dual results are summarized in Table 3.8.

$\overset{\pi}{\neg}\left(\sigma \wedge \overset{\pi}{\neg}\sigma\right) = \mathbf{1}_U$	$\overset{\pi}{-}\left(\sigma \vee \overset{\pi}{-}\sigma\right) = \mathbf{0}_U$
$\overset{\pi}{\neg}\sigma = \sigma \Rightarrow \pi = \mathbf{1}_U$ iff $\sigma \precsim \pi$	$\overset{\pi}{-}\sigma = \pi - \sigma = \mathbf{0}_U$ iff $\pi \precsim \sigma$
$(\pi - \sigma) \Rightarrow \pi = \overset{\pi}{\neg}\overset{\pi}{-}\sigma = \mathbf{1}_U$	$\pi - (\sigma \Rightarrow \pi) = \overset{\pi}{-}\overset{\pi}{\neg}\sigma = \mathbf{0}_U$
$\pi \wedge \sigma = (\sigma \Rightarrow \pi) \wedge \sigma$	$\pi \vee \sigma = (\pi - \sigma) \vee \sigma$
$\neg(\sigma \vee \pi) = \neg\sigma \wedge \neg\pi$	$-(\sigma \wedge \pi) = -\sigma \vee -\pi$

Table 3.8: Some duality results

Chapter 4

The quantum logic of vector-space partitions (DSDs)

4.1 Linearization from sets to vector spaces

The logic of (set) partitions is dual to the usual logic of subsets (usually presented in the special case of propositional logic) in the sense of the category-theoretic duality between partitions and subsets.[1] The usual quantum logic [11] first generalizes from the set notion of a subset to the corresponding vector-space notion of a subspace and then specializes the vector spaces to those used in quantum mechanics. Our purpose is to present the dual quantum logic by first generalizing the set notion of a partition to the corresponding vector-space notion, and then specialize to the (finite-dimensional) vector spaces of quantum mechanics. But what is the vector-space notion corresponding to the set notion of partition?

There is a 'semi-algorithmic' procedure to generalize set concepts to the corresponding vector-space notions that we will call *linearization*. Gian-Carlo Rota might call this procedure a "yoga" [91, p. 251]. The basic idea is simple.

Yoga of Linearization: Take a basis set of a vector space V (over a field $\mathbb{K}$) and apply to that basis set any set notion, and then whatever is generated

---

[1] As Gian-Carlo Rota put it: "categorically speaking, the Boolean σ-algebra of events and the lattice Σ of all Boolean σ-subalgebras are dual notions" [92, p. 65] using the characterization of partitions by Boolean subalgbras [66, p. 43] that goes back to Ore [79]. The category theorist, F. William Lawvere, called subobjects "parts" and then noted that: "The dual notion (obtained by reversing the arrows) of 'part' is the notion of partition." [69, p. 85]

in the vector space is the corresponding vector-space notion.

The set notion of cardinality applied to a basis set generates the notion of dimension. The set notion of a subset applied to a basis set generates the notion of a subspace. The set notion of a partition applied to a basis set generates the notion of a direct-sum decomposition. If the set partition of the basis set is the inverse image of a numerical attribute[2] $f : U \to \mathbb{K}$, then it generates the notion of a (diagonalizable = there is a basis set of eigenvectors) linear operator $F : V \to V$ defined by $Fu = f(u)u$ linearly extended to the whole space. For $r \in \mathbb{K}$ in the image $f(U)$ is a subset of $\mathbb{K}$, the inverse image $f^{-1}(r)$ is a constant subset of U. If for $S \subseteq U$, we let rS stand for defining a function having the value r on the elements of S, then the numerical attribute $f : U \to \mathbb{K}$ satisfies $f(f^{-1}(r)) = rf^{-1}(r)$ and similarly for any subset $S \subseteq f^{-1}(r)$, $f(S) = rS$. Then the equation characterizing the constant sets of f is $f(S) = rS$ for some $r \in \mathbb{K}$, and the vector-space version is the eigenvector equation $Fv = \lambda v$ for some $\lambda \in \mathbb{K}$. The set of constant r-sets of f is the powerset $\wp(f^{-1}(r))$ for some $r \in f(U)$, and the corresponding space of eigenvectors of F for some eigenvalue λ is the eigenspace V_λ associated with λ. Starting with a diagonalizable linear operator $F : V \to V$, the numerical attribute $f : U \to \mathbb{K}$ reappears as the eigenvalue function on a basis of eigenvectors. Given two vector spaces V and V' over $\mathbb{K}$ with basis sets U and U', the set notion of the direct or Cartesian product $U \times U'$ generates the notion of the tensor product $V \otimes V'$ generated by the ordered pairs $(u, u') \in U \times U'$ written as $u \otimes u'$. These results are summarized in Table 4.1.[3]

---

[2] A attribute or property might be said to hold or not hold of an element in a set, but a numerical attribute, like height, weight, or age, assigns a numerical value to each element of a set. And an attribute can be considered as a numerical attribute with possible values $0, 1 \in 2$.

[3] Since set-concepts can be formulated in vector spaces over $\mathbb{Z}_2$, the vector-space concepts of QM over $\mathbb{C}$ can be scaled back to $\mathbb{Z}_2$ in an appropriate way to create a pedagogical or 'toy' model of QM over $\mathbb{Z}_2$, i.e., quantum mechanics over sets [26].

Set concept	Vector-space concept
Universe set U	Basis set of a space V
Cardinality of a set U	Dimension of a space V
Subset of a set U	Subspace of a space V
Partition of a set U	Direct-sum decomposition of a space V
Numerical attribute $f : U \to \mathbb{K}$	Diagonalizable linear op. $F : V \to V$
Value r in image $f(U)$ of f	Eigenvalue λ of F
Constant set S of f	Eigenvector v of F
Set of constant r-sets $\wp\left(f^{-1}(r)\right)$	Eigenspace V_λ of λ
Direct product of sets	Tensor product of spaces
Elements (u, u') of $U \times U'$	Basis vectors $u \otimes u'$ of $V \otimes V'$

Table 4.1: Linearization of set concepts to corresponding vector-space concepts.

The usual notion of logic is the Boolean logic of subsets (ordinarily represented in the special case of propositional logic) and since subsets linearize to subspaces, the usual quantum logic [11] is the logic of subspaces of a vector space specialized to the Hilbert spaces of quantum mechanics. Since set partitions are category-theoretic duals to subsets, we have presented the dual logic of set partitions above, and the corresponding quantum logic is the logic of direct-sum decompositions [28] of a vector space specialized to the Hilbert spaces of quantum mechanics. The general logic of direct-sum decompositions (i.e., vector space partitions) is the subject of this chapter with some specializations to the quantum case.

4.2 Direct-Sum Decompositions

Let V be a finite-dimensional vector space over a field $\mathbb{K}$. Intuitively, a *direct-sum decomposition (DSD)* of V is a set of non-zero subspaces $\{V_i\}_{i \in I}$ of V such that every vector $v \in V$ can be uniquely expressed as a sum $v = \sum_{i \in I} v_i$ where $v_i \in V_i$. We have seen how the notion of a set partition linearizes to the notion of a DSD when a set partition on a set U is usually defined as a set of non-empty subsets or blocks $\{B_i\}_{i \in I}$ such that the blocks are mutually disjoint and jointly exhaustive (i.e., whose union is U). But it might be noted that a set partition could be equivalently defined as a 'delinearized' DSD. That is, a set partition is a set of non-empty subsets $\{B_i\}_{i \in I}$ of U such that every non-empty subset $S \subseteq U$ can be uniquely expressed as the union of subsets of the B_i (e.g., the projections $S \cap B_i$) for $i \in I$. For a spectrum of values $r_1, ..., r_m$, for $f : U \to \mathbb{R}$ with $B_i = f^{-1}(r_i)$, the unique decomposition of $S \subseteq U$ is illustrated in Figure

4.1.

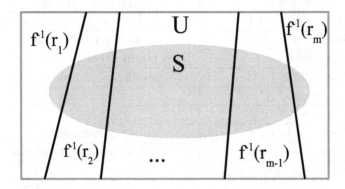

Figure 4.1: Decomposition of $S \subseteq U$ in terms of the constant sets of
$f : U \to \mathbb{R}$.

Every self-adjoint (or Hermitian) operator in a Hilbert space, or, more generally, any diagonalizable operator $V \to V$ determines the DSD of its eigenspaces
V_i consisting of the eigenvectors v_i for an eigenvector λ_i. But the notion of
a DSD makes sense independently of diagonalizable linear operators and their
eigenspaces. For instance, in the pedagogical model of "QM over sets" [26], the
vector space is $\mathbb{Z}_2^n$, and the only linear operators $\mathbb{Z}_2^n \to \mathbb{Z}_2^n$ are the projection
operators to a subspace and their DSDs of eigenspaces are always binary corre-
sponding to the two possible eigenvalues of $1, 0 \in \mathbb{Z}_2$. But there can be DSDs of
any cardinality ($\leq n$) determined by (say) real-valued numerical attributes on
a basis set. Let $U = \{u_1, ..., u_n\}$ be a basis set for $\mathbb{Z}_2^n$ (so each vector in $\mathbb{Z}_2^n$ 'is'
just a subset $S \subseteq U$) and let $f : U \to \mathbb{R}$ be a numerical attribute on U (where
the values of the attribute are not in the base field $\mathbb{Z}_2$). Then there is the DSD
of subspaces $\wp\left(f^{-1}(r)\right)$ of the constant sets (or 'eigenvectors') $f(S) = rS$ for
$S \subseteq U$ and r a value (or 'eigenvalue') in the image (or 'spectrum') $f(U) \subseteq \mathbb{R}$.
As the base field is increased up to the complex numbers, then all real-valued
observables can be "internalized" as self-adjoint operators taking values in the
base field.

Let $DSD(V)$ be the set of DSDs on the vector space V. It is also convenient
to think of DSDs as just vector-space partitions with the subspaces V_i as the
blocks in the partition. Then it is easily seen that there is a natural refinement
partial order on the vector-space partitions on a vector space V, That is, if
$\pi = \{V_i\}_{i \in I}$ and $\sigma = \{W_j\}_{j \in J}$, then σ is *refined* by π, $\sigma \precsim \pi$, if for every
$V_i \in \pi$, there is a $W_j \in \sigma$ such that $V_i \subseteq W_j$–so that $DSD(V)$ becomes a
partially ordered set. The partially ordered set has a bottom element, namely

the *indiscrete* DSD $\mathbf{0} = \{V\}$ (nicknamed 'the blob')–but no unique top element since each different basis set will determine a different DSD of one-dimensional subspaces generated by the basis set.

4.3 Compatibility of DSDs

For our proofs, a more rigorous definition of a DSD or vector-space partition is:

Definition 92 *Let V be a finite dimensional vector space over a field $\mathbb{K}$. A direct sum decomposition (DSD) of V is a set of subspaces $\{V_i\}_{i \in I}$ such that $V_i \cap \sum_{i' \neq i} V_{i'} = \{0\}$ (the zero space) for $i \in I$ and which span the space, i.e., in terms of direct sums, $\oplus_{i \in I} V_i = V$.*

To fix notation, the following are arbitrary DSDs: $\pi = \{V_i\}_{i \in I}$, $\sigma = \{W_j\}_{j \in J}$, and $\tau = \{X_k\}_{k \in K}$ of V.

In the logic of (set) partitions, we had the luxury of assuming a fixed universe set U, but with vector-space partitions, there are many different basis sets so we have to deal with the problem of incompatibility (or non-commutativity) of DSDs. When the DSDs are generated by linear operators, this is the non-commutativity of operators, but we are dealing with DSD's without assuming any corresponding operators. Intuitively, a diagonalizable linear operator is determined by a DSD plus an eigenvalue in the base field for each distinct subspace in the DSD. But the most important thing about eigenvalues is not their actual values but when they are the same or different, and that information is conveyed by the DSDs. Hence we need to define compatibility and incompatibility between DSDs to correspond to the usual distinction between commutativity and non-commutativity of operators.

Given two DSDs $\pi = \{V_i\}_{i \in I}$ and $\sigma = \{W_j\}_{j \in J}$, their *proto-join* $\pi \vee \sigma$ is the set of non-zero subspaces $\{V_i \cap W_j | V_i \cap W_j \neq \{0\}\}_{(i,j) \in I \times J}$. The proto-join $\pi \vee \sigma$ is not necessarily a DSD. The space spanned by the proto-join is denoted $\mathcal{SE}$. If the two DSDs π and σ were defined as the eigenspace DSDs of two diagonalizable operators, then the space $\mathcal{SE}$ spanned by the proto-join would be the space spanned by the simultaneous eigenvectors of the two operators (and hence the notation $\mathcal{SE}$).

Proposition 16 *Let $F, G : V \to V$ be two diagonalizable linear operators on a finite dimensional vector space V. Then $\mathcal{SE}$ is the kernel of the commutator: $\mathcal{SE} = \ker([F, G]) = \ker(FG - GF)$.*

Proof: Let v be any simultaneous eigenvector of the operators, i.e., $Fv = \lambda v$ and $Gv = \mu v$. Then $[F, G](v) = (FG - GF)(v) = (\lambda \mu - \mu \lambda) v = 0$ so the space $\mathcal{SE}$ spanned by the simultaneous eigenvectors is contained in the kernel ker $([F,G])$, i.e., $\mathcal{SE} \subseteq \ker([F,G])$. Conversely, if we restrict the two operators to the subspace ker $([F,G])$, then the restricted operators commute on that subspace. Then it is a standard theorem of linear algebra [54, p. 177] that the subspace ker $([F,G])$ is spanned by simultaneous eigenvectors of the two restricted operators. But if a vector is a simultaneous eigenvector for the two operators restricted to a subspace, they are the same for the operators on the whole space V, since the two conditions $Fv = \lambda v$ and $Gv = \mu v$ only involves a vector in the subspace. Hence ker $([F,G]) \subseteq \mathcal{SE}$. $\square$

The operators commute if the commutator $[F,G]$ is the zero operator, i.e., $[F,G]v = FG - GF = 0$ for any $v \in V$, so the kernel of the commutator is the whole space V. The commutativity definitions for two DSDs π and σ without using operators are:

Definition 93 π, σ *commute or are compatible, denoted* $\pi \leftrightarrow \sigma$, *if their proto-join* $\pi \vee \sigma$ *spans the whole space, i.e.,* $\mathcal{SE} = V$.

When two DSDs π and σ are compatible, $\pi \leftrightarrow \sigma$, their proto-join is the *join DSD* :

$$\pi \vee \sigma = \{V_i \cap W_j | V_i \cap W_j \neq \{0\}\}_{(i,j) \in I \times J}$$

Join of DSDs when $\pi \leftrightarrow \sigma$.

When the proto-join $\pi \vee \sigma$ is the join so that $\pi \vee \sigma$ is a DSD, then it is clear from the refinement partial ordering of DSDs that $\pi \vee \sigma$ is the join in the sense of the least upper bound on π and σ. Since the subspaces in the join of any DSD π with the indiscrete DSD **0** are just the subspaces of π, their proto-join is a join and:

$$\pi \vee \mathbf{0} = \pi.$$

Moreover this means that the indiscrete DSD is compatible with all DSDs, i.e., for any DSD π:

$$\mathbf{0} \leftrightarrow \pi.$$

A set of partitions on a set U might be said to be complete if their join is the discrete partition $\mathbf{1}_U$ whose blocks are singletons. Similarly, in a vector space, a set of compatible (or commuting) DSDs is said to be *complete* is their join is a DSD of one dimensional subspaces. In terms of operators, this is Dirac's Complete Set of Commuting Observables or CSCO [21]. Then the simultaneous 'eigenvector' determined by each one-dimensional intersection can be characterized by the ordered sequence of 'eigenvalues' it had in each DSD of the complete set when the subspaces in each of the DSDs are differentiated by different numbers as if they were the eigenspaces of commuting observables.

Compatibility defines a binary relation on $DSD(V)$ that is clearly reflexive and symmetric. The relation cannot be transitive since $\mathbf{0}$ is compatible with all DSDs and then $\pi \leftrightarrow \mathbf{0} \leftrightarrow \sigma$ would imply that all DSDs are compatible (see below for counterexamples)–so compatibility is not an equivalence relation. The join operation preserves compatibility, i.e., if $\pi \leftrightarrow \sigma$, then $\pi \leftrightarrow \pi \vee \sigma \leftrightarrow \sigma$. The important thing is that if a set of three DSDs are mutually compatible, then their three-way join is a DSD. To prove that, we need a Lemma.

Lemma 94 *Let the DSDs $\pi = \{V_i\}_{i \in I}$ and $\sigma = \{W_j\}_{j \in J}$ be compatible so that $\pi \vee \sigma$ is a DSD and thus any $v \in V$ has a unique expression $v = \sum_{(i,j) \in I \times J} v_{ij}$ where $v_{ij} \in V_i \cap W_j$. Let $v_i = \sum_{j \in J} v_{ij}$ so that $v_i \in V_i$ and then $v = \sum_{i \in I} v_i$. If $v \in V_i$, then $v = v_i$.*

Proof: Let $\widehat{v_i} = \sum_{i' \in I, i' \neq i} v_{i'}$ so that $v = v_i + \widehat{v_i}$. Hence if $v \in V_i$, then $v - v_i = \widehat{v_i} \in V_i$. Since $\widehat{v_i} \in \oplus_{i' \in I, i' \neq i} V_{i'}$, $\widehat{v_i} \in V_i \cap \oplus_{i' \in I, i' \neq i} V_{i'}$ so $\widehat{v_i} = 0$ since $\pi = \{V_i\}_{i \in I}$ is a DSD which implies $V_i \cap \oplus_{i' \in I, i' \neq i} V_{i'} = \{0\}$. If $\widehat{v_i} = 0$, then $v = v_i$. $\qquad \square$

Theorem 95 *Given three DSDs, $\pi = \{V_i\}_{i \in I}$, $\sigma = \{W_j\}_{j \in J}$, and $\tau = \{X_k\}_{k \in K}$ that are mutually compatible, i.e., $\pi \leftrightarrow \sigma$, $\pi \leftrightarrow \tau$, and $\sigma \leftrightarrow \tau$, then $(\pi \vee \sigma) \leftrightarrow \tau$ and equivalently , $\pi \leftrightarrow \sigma \vee \tau$ and thus $\pi \vee \sigma \vee \tau$ is a DSD.*

Proof: We need to prove $\pi \vee \sigma \leftrightarrow \tau$ or equivalently $\pi \leftrightarrow \sigma \vee \tau$, i.e., that

$$\oplus_{(i,j,k) \in I \times J \times K} ((V_i \cap W_j) \cap X_k) = \oplus_{(i,j,k) \in I \times J \times K} (V_i \cap (W_j \cap X_k)) = V.$$

Consider any nonzero $v \in V$ where since $\pi \leftrightarrow \sigma$, there are $v_{ij} \in V_i \cap W_j$ for each $i \in I$ and $j \in J$ such that $v = \sum_{(i,j) \in I \times J} v_{ij}$. Consider any such nonzero v_{ij}. Now since $\pi \leftrightarrow \tau$, there are $v_{ij,i'k} \in V_{i'} \cap X_k$ for each $i' \in I$ and $k \in K$ such that $v_{ij} = \sum_{(i',k) \in I \times K} v_{ij,i'k}$. But since $v_{ij} \in V_i$, by the Lemma, only $v_{ij,ik}$ is nonzero, so $v_{ij} = \sum_{k \in K} v_{ij,ik}$. Symmetrically, since $\sigma \leftrightarrow \tau$, there are $v_{ij,j'k} \in W_{j'} \cap X_k$ for each $j' \in J$ and $k \in K$ such that $v_{ij} = \sum_{(j',k) \in J \times K} v_{ij,j'k}$. But since $v_{ij} \in W_j$, by the Lemma, only $v_{ij,jk}$ is nonzero, so $v_{ij} = \sum_{k \in K} v_{ij,jk}$. Now since $\{X_k\}_{k \in K}$ is a DSD, there is a unique expression for $v_{ij} = \sum_{k \in K} v_{ijk}$ where $v_{ijk} \in X_k$. Hence by uniqueness: $v_{ijk} = v_{ij,ik} = v_{ij,jk}$. But since $v_{ij,ik} \in V_i$ and $v_{ij,jk} \in W_j$ and $v_{ij,ik} = v_{ijk} = v_{ij,jk}$, we have $v_{ijk} \in V_i \cap W_j \cap X_k$. Thus $v = \sum_{(i,j) \in I \times J} v_{ij} = \sum_{(i,j) \in I \times J} \sum_{k \in K} v_{ijk} = \sum_{(i,j,k) \in I \times J \times K} v_{ijk}$. Since v was arbitrary, $\oplus_{(i,j,k) \in I \times J \times K} (V_i \cap W_j \cap X_k) = V$, so $\pi \vee \sigma \vee \tau$ is a DSD. $\qquad \square$

Definition 96 π, σ *are incompatible if $\mathcal{SE} \neq V$; and*

Definition 97 π, σ *are conjugate if $\mathcal{SE} = \{\emptyset\}$.*

4.4 Examples of compatibility, incompatibility, and conjugacy

Example of incompatibility (or non-commutativity). To illustrate these concepts, it suffices to take the simplest case of a vector space over $\mathbb{Z}_2$. Consider the $2^3 = 8$-element vector space $\mathbb{Z}_2^3$ with a standard computational basis $\{e_1, e_2, e_3\}$ where the vectors may be represented as column vectors;

$$e_1 = \begin{bmatrix} 1 \\ 0 \\ 0 \end{bmatrix}, \; e_2 = \begin{bmatrix} 0 \\ 1 \\ 0 \end{bmatrix}, \text{ and } e_3 = \begin{bmatrix} 0 \\ 0 \\ 1 \end{bmatrix}.$$

An alternative basis is:

$$a_1 = \begin{bmatrix} 1 \\ 1 \\ 0 \end{bmatrix}, \; a_2 = \begin{bmatrix} 0 \\ 1 \\ 1 \end{bmatrix}, \text{ and } a_3 = \begin{bmatrix} 1 \\ 1 \\ 1 \end{bmatrix}.$$

The a_i form a basis since $a_1 + a_3 = e_3$ (remember that $1 + 1 = 0$ in $\mathbb{Z}_2$), $a_2 + a_3 = e_1$, and $a_1 + a_2 + a_3 = e_2$. To define a DSD, all we need is a numerical attribute defined on a basis set. Consider $f : U = \{e_1, e_2, e_3\} \rightarrow \mathbb{R}$ defined by: $f(e_1) = 1$ and $f(e_2) = f(e_3) = 17$ (the values don't matter; all that matters is when they are the same or different). Let $g : U' = \{a_1, a_2, a_3\} \rightarrow \mathbb{R}$ be defined by: $g(a_1) = g(a_2) = 2$ and $g(a_3) = 5$. Then f defines a DSD π with two blocks: $f^{-1}(1) = \{e_1\}$ which generates the subspace $[f^{-1}(1)] = \{0, e_1\}$ (where the square brackets $[S]$ indicate the subspace generated by the vectors in S) and $f^{-1}(17) = \{e_2, e_3\}$ with generates the subspace $[f^{-1}(17)] = \{0, e_2, e_3, e_2 + e_3\}$. Note that in terms of subset of the basis set $\{e_1, e_2, e_3\}$, the subspaces generated by the inverse images are just the powersets: $\wp(f^{-1}(1)) = \{\emptyset, \{e_1\}\}$ and $\wp(f^{-1}(17)) = \{\emptyset, \{e_2\}, \{e_3\}, \{e_2, e_3\}\}$. The other numerical attribute g defines another DSD σ with two blocks: $g^{-1}(2) = \{a_1, a_2\}$ which generates the subspace $[g^{-1}(2)] = \{0, a_1, a_2, a_1 + a_2\}$ and $g^{-1}(5) = \{a_3\}$ which generates the subspace $[g^{-1}(5)] = \{0, a_3\}$. Then the direct-sums are the whole space: $[f^{-1}(1)] \oplus [f^{-1}(17)] = \mathbb{Z}_2^3 = [g^{-1}(2)] \oplus [g^{-1}(5)]$. To compute the proto-join of the two DSDs, we need to express the vectors spaces in the same (computational) basis so: $[g^{-1}(2)] = \{0, a_1, a_2, a_1 + a_2\} = \{0, e_1 + e_2, e_2 + e_3, e_1 + e_3\}$ and $[g^{-1}(5)] = \{0, a_3\} = \{0, e_1 + e_2 + e_3\}$. Since there are two blocks in each DSD π and σ, there are four possible non-zero intersections in the proto-join $\pi \vee \sigma$, but the only non-zero intersection is the subspace generated by the intersection of $[f^{-1}(17)]$ and $[g^{-1}(2)]$:

$$\left[f^{-1}(17)\right] \cap \left[g^{-1}(2)\right] = \{0, e_2, e_3, e_2 + e_3\} \cap \{0, e_1 + e_2, e_2 + e_3, e_1 + e_3\} =$$
$$\{0, e_2 + e_3\}.$$

Since the simultaneous 'eigenvector' space $\mathcal{SE} = \{0, e_2 + e_3\}$ is neither $V = \mathbb{Z}_2^3$ nor the zero space, π and σ are incompatible but not conjugate.

Example of compatibility. Let U and f be the same as above but consider the basis: $b_1 = e_1$, $b_2 = e_2$, and $b_3 = e_2 + e_3$. Let $g : U'' = \{b_1, b_2, b_3\} \to \mathbb{R}$ where $g(b_1) = 1$, $g(b_2) = 2$, and $g(b_3) = 3$ so the three subspaces in the DSD σ are $\left[g^{-1}(1)\right] = \{0, b_1\} = \{0, e_1\}$, $\left[g^{-1}(2)\right] = \{0, b_2\} = \{0, e_2\}$, and $\left[g^{-1}(3)\right] = \{0, b_3\} = \{0, e_2 + e_3\}$. There are six subspaces determined by the intersections of the two subspaces given by f and the three given by g, but the only non-zero intersections are:

$$\{0, e_1\}, \{0, e_2\}, \text{ and } \{0, e_2 + e_3\}.$$

These three subspaces span the whole space so $\mathcal{SE} = V$, and π and σ are compatible. Each of these 'eigenspaces' is one-dimensional so the DSDs π and σ given by f and g are a complete set of DSDs and the simultaneous 'eigenvectors' in the one-dimensional blocks can be characterized by the sequence of 'eigenvalues', i.e., $e_1 = |1, 1\rangle$, $e_2 = |17, 2\rangle$, and $e_2 + e_3 = |17, 3\rangle$, when the DSDs are determined by numerical attributes to attach 'eigenvalues' to the subspaces in the DSDs.

Example of conjugacy. For a standard example of conjugacy, we need to work in $\mathbb{Z}_2^n$ where n is even so consider $\mathbb{Z}_2^4$ with the computational basis $U = \{e_1, ..., e_4\}$. Let $\hat{U} = \{\hat{e}_1, ..., \hat{e}_4\}$ where $\hat{e}_i$ is the vector formed by excluding e_i from the sum $\sum_{j=1}^{4} e_j$. These four vectors form a basis set for $\mathbb{Z}_2^4$–while the corresponding set of vectors for n odd do not form a basis. Let $f : U \to \mathbb{R}$ be defined by $f(e_1) = f(e_2) = 1$ and $f(e_3) = f(e_4) = 0$ so the two subspaces in π are $\left[f^{-1}(1)\right] = \{0, e_1, e_2, e_1 + e_2\}$ and $\left[f^{-1}(0)\right] = \{0, e_3, e_4, e_3 + e_4\}$. Let $g : \hat{U} \to \mathbb{R}$ be given by $g(\hat{e}_2) = g(\hat{e}_3) = 1$ and $g(\hat{e}_1) = g(\hat{e}_4) = 1$ so that the two subspaces in σ are:

$$\begin{aligned}
\left[g^{-1}(1)\right] &= \{0, \hat{e}_2, \hat{e}_3, \hat{e}_2 + \hat{e}_3\} = \{0, e_1 + e_3 + e_4, e_1 + e_2 + e_4, e_2 + e_3\}, \\
\left[g^{-1}(0)\right] &= \{0, \hat{e}_1, \hat{e}_4, \hat{e}_1 + \hat{e}_4\} = \{0, e_2 + e_3 + e_4, e_1 + e_2 + e_3, e_1 + e_4\}.
\end{aligned}$$

Then it is easily checked that the four intersections of subspaces in the proto-join $\pi \vee \sigma$ are all the zero space $\{0\}$, e.g., f and g have no simultaneous 'eigenvectors', so π and σ are conjugate.

Since the two numerical attributes in the conjugacy example are characteristic functions taking value in the base field $\mathbb{Z}_2$, they define linear operators F

and G on $\mathbb{Z}_2^4$ and thus we can compute their commutator $[F, G] = FG - GF$ as usual once restated in the computational basis. The matrix to convert a $0, 1$-vector written in the $\hat{U}$-basis to the same $0, 1$-vector written in the U-basis is:

$$C_{U \leftarrow \hat{U}} = \begin{bmatrix} 0 & 1 & 1 & 1 \\ 1 & 0 & 1 & 1 \\ 1 & 1 & 0 & 1 \\ 1 & 1 & 1 & 0 \end{bmatrix}.$$

For instance, $e_1 = \hat{e}_2 + \hat{e}_3 + \hat{e}_4$ so that vector in $\hat{U}$-basis is the column vector $[0, 1, 1, 1]^t$ (t represents the transpose) and

$$\begin{bmatrix} 0 & 1 & 1 & 1 \\ 1 & 0 & 1 & 1 \\ 1 & 1 & 0 & 1 \\ 1 & 1 & 1 & 0 \end{bmatrix} \begin{bmatrix} 0 \\ 1 \\ 1 \\ 1 \end{bmatrix} \overset{\mathrm{mod}(2)}{=} \begin{bmatrix} 1 \\ 0 \\ 0 \\ 0 \end{bmatrix}$$

which is e_1 in the U-basis. The inverse to the conversion matrix $C_{U \leftarrow \hat{U}}$ is:

$$C_{\hat{U} \leftarrow U} = \begin{bmatrix} 0 & 1 & 1 & 1 \\ 1 & 0 & 1 & 1 \\ 1 & 1 & 0 & 1 \\ 1 & 1 & 1 & 0 \end{bmatrix}$$

so the conversion matrix is its own inverse. The numerical attribute f now defines a linear operator $F : \mathbb{Z}_2^4 \to \mathbb{Z}_2^4$ given by $Fe_1 = 1e_1$ land $Fe_2 = 1e_2$ as well as $Fe_3 = 0e_3$ and $Fe_4 = 0e_4$ both of which are the zero vector 0. The numerical attribute g also defines a linear operator $G : \mathbb{Z}_2^4 \to \mathbb{Z}_2^4$ given by $G\hat{e}_2 = 1\hat{e}_2$ and $G\hat{e}_3 = 1\hat{e}_3$ as well as $G\hat{e}_1 = 0\hat{e}_1$ and $G\hat{e}_4 = 0\hat{e}_4$ both of which are the vector 0. We need to convert the matrix for the operator G defined in the $\hat{U}$-basis into the U-basis in order to commute the commutator of the two operators. The matrix representation of the G operator in the $\hat{U}$-basis is just the projection operator P_1 to the subspace $[g^{-1}(1)] = \{0, \hat{e}_2, \hat{e}_3, \hat{e}_2 + \hat{e}_3\}$ associated with the eigenvalue 1. The conversion of that matrix to the U-basis is accomplished by pre- and post- multiplying by the appropriate conversion matrices:

$$C_{U \leftarrow \hat{U}} P_1 C_{\hat{U} \leftarrow U}$$

$$= \begin{bmatrix} 0 & 1 & 1 & 1 \\ 1 & 0 & 1 & 1 \\ 1 & 1 & 0 & 1 \\ 1 & 1 & 1 & 0 \end{bmatrix} \begin{bmatrix} 0 & 0 & 0 & 0 \\ 0 & 1 & 0 & 0 \\ 0 & 0 & 1 & 0 \\ 0 & 0 & 0 & 0 \end{bmatrix} \begin{bmatrix} 0 & 1 & 1 & 1 \\ 1 & 0 & 1 & 1 \\ 1 & 1 & 0 & 1 \\ 1 & 1 & 1 & 0 \end{bmatrix}$$

$$= \begin{bmatrix} 2 & 1 & 1 & 2 \\ 1 & 1 & 0 & 1 \\ 1 & 0 & 1 & 1 \\ 2 & 1 & 1 & 2 \end{bmatrix} \overset{\mathrm{mod}(2)}{=} \begin{bmatrix} 0 & 1 & 1 & 0 \\ 1 & 1 & 0 & 1 \\ 1 & 0 & 1 & 1 \\ 0 & 1 & 1 & 0 \end{bmatrix}.$$

The matrix representing the $F : \mathbb{Z}_2^4 \to \mathbb{Z}_2^4$ in the U-basis is the projection matrix:

$$\begin{bmatrix} 1 & 0 & 0 & 0 \\ 0 & 1 & 0 & 0 \\ 0 & 0 & 0 & 0 \\ 0 & 0 & 0 & 0 \end{bmatrix}$$

and the commutator in terms of the operators is $[F, G] = FG - GF$, so it is computed in the U-basis as:

$$\begin{bmatrix} 1 & 0 & 0 & 0 \\ 0 & 1 & 0 & 0 \\ 0 & 0 & 0 & 0 \\ 0 & 0 & 0 & 0 \end{bmatrix} \begin{bmatrix} 0 & 1 & 1 & 0 \\ 1 & 1 & 0 & 1 \\ 1 & 0 & 1 & 1 \\ 0 & 1 & 1 & 0 \end{bmatrix} - \begin{bmatrix} 0 & 1 & 1 & 0 \\ 1 & 1 & 0 & 1 \\ 1 & 0 & 1 & 1 \\ 0 & 1 & 1 & 0 \end{bmatrix} \begin{bmatrix} 1 & 0 & 0 & 0 \\ 0 & 1 & 0 & 0 \\ 0 & 0 & 0 & 0 \\ 0 & 0 & 0 & 0 \end{bmatrix}$$

$$= \begin{bmatrix} 0 & 0 & 1 & 0 \\ 0 & 0 & 0 & 1 \\ -1 & 0 & 0 & 0 \\ 0 & -1 & 0 & 0 \end{bmatrix} \overset{\mathrm{mod}(2)}{=} \begin{bmatrix} 0 & 0 & 1 & 0 \\ 0 & 0 & 0 & 1 \\ 1 & 0 & 0 & 0 \\ 0 & 1 & 0 & 0 \end{bmatrix}.$$

The resulting matrix, representing the commutator operator $FG - GF$, has a determinant of 1 so it is a non-singular transformation with a kernel of the zero space $\{0\}$. Hence the definition of π and σ being conjugate agrees with the kernel of the commutator $[F, G] = FG - GF : \mathbb{Z}_2^4 \to \mathbb{Z}_2^4$ being the zero space when the numerical attributes f and g define operators $F, G : \mathbb{Z}_2^4 \to \mathbb{Z}_2^4$. All these examples illustrate general aspects of DSDs with no restriction to Hilbert spaces. In fact, by using vector spaces over $\mathbb{Z}_2$ instead of $\mathbb{C}$, all the examples are from the pedagogical model of "quantum mechanics over sets" [26] since vectors in $\mathbb{Z}_2^n$ can be interpreted as subsets of an n-element set, and the different n-element basis sets allows the introduction of the notions on commutativity and non-commutativity that are important in QM over $\mathbb{C}$.

4.5   The meet of DSDs and properties of refinement

The meet of two set partitions $\pi = \{B_i\}_{i \in I}$ and $\sigma = \{C_j\}_{j \in J}$ was defined as the set partition $\pi \wedge \sigma$ whose blocks are a union of some π-blocks and also a union of some σ-blocks, and those blocks are minimal in that regard. The meet of two vector-space partitions or DSDs is just the vector space version of the set definition

Definition 98 *For any two DSDs* $\pi = \{V_i\}_{i \in I}$ *and* $\sigma = \{W_j\}_{j \in J}$, *the meet* $\pi \wedge \sigma$ *is the DSD whose subspaces are direct sums of subspaces from* π *and the direct sum of subspaces from* σ *and are minimal subspaces in that regard. That is,* $\{Y_l\}_{l \in L}$ *is the meet if there is a set partition* $\{I_l\}_{l \in L}$ *on* I *and a set partition* $\{J_l\}_{l \in L}$ *on* J *such that for all* $l \in L$. $Y_l = \oplus_{i \in I_l} V_i = \oplus_{j \in J_l} W_j$ *and that holds for no more refined partitions on the index sets.*

Note that for the blob $\mathbf{0} = \{V\}$, $V = \oplus_{i \in I} V_i = \oplus_{j \in J} W_j$ using the blob set partitions $\{I\}$ and $\{J\}$, but in general the meet $\pi \wedge \sigma$ will use more refined partitions on I and J. Thus $\mathbf{0}$ is always a lower bound on π and σ, and we show below that $\pi \wedge \sigma$ is the least such lower bound.

Proposition 17 *If* $\pi \leftrightarrow \tau$ *or* $\sigma \leftrightarrow \tau$, *then* $\pi \wedge \sigma \leftrightarrow \tau$.

Proof: Suppose that $\pi \leftrightarrow \tau$ so that $\{V_i \cap X_k\}_{(i,k) \in I \times K}$ is a DSD, i.e., $\oplus_{(i,k) \in I \times K} \{V_i \cap X_k\} = V$. The blocks of $\pi \wedge \sigma$ for $Y_l = \oplus_{i \in I_l} V_i = \oplus_{j \in J_l} W_j$. To show that $\pi \wedge \sigma \leftrightarrow \tau$, we need to show that $\{Y_l \cap X_k\}_{(l,k) \in L \times K}$ spans the whole space V, i.e., $\oplus_{(l,k) \in L \times K} Y_l \cap X_k = V$. Now $Y_l \cap X_k = (\oplus_{i \in I_l} V_i) \cap X_k$ and since $\pi \leftrightarrow \tau$, $\{V_i \cap X_k\}_{(i,k) \in I \times K}$ is a DSD, so $V_i = \oplus_{k \in K} V_i \cap X_k$ and then $Y_l \cap X_k = (\oplus_{i \in I_l} V_i) \cap X_k = (\oplus_{(i,k') \in I_l \times K} V_i \cap X_{k'}) \cap X_k$ [56, p. 30]. But the intersection between $(\oplus_{(i,k') \in I_l \times K} V_i \cap X_{k'})$ and X_k is $V_i \cap X_k$ since no non-zero vector in $(\oplus_{(i,k') \in I_l \times K, k' \neq k} V_i \cap X_{k'})$ can be in X_k. Thus $Y_l \cap X_k = \oplus_{i \in I_l} V_i \cap X_k$ and

$$\oplus_{(l,k) \in L \times K} Y_l \cap X_k = \oplus_{(l,k) \in L \times K} \oplus_{i \in I_l} V_i \cap X_k = \oplus_{(i,k) \in I \times K} \{V_i \cap X_k\} = V$$

which is a DSD since $\pi \leftrightarrow \tau$. If $\sigma \leftrightarrow \tau$, the proof is symmetrical. $\square$

"The Blob" absorbs everything it meets:

$$\mathbf{0} \wedge \pi = \mathbf{0}.$$

It may be recalled that the refinement partial order on DSDs is defined just like the refinement partial order on partitions but with the subspaces of the DSD replacing the subsets or blocks of the partition. In the partition refinement $\sigma \precsim \pi$, we have for each block $C \in \sigma$, $C = \cup \{B \in \pi : B \subseteq C\}$. The same holds for vector-space partitions.

Lemma 99 *If $\sigma \preceq \pi$, then each subspace $W_j \in \sigma$, $W_j = \oplus \{V_i \in \pi : V_i \subseteq W_j\}$.*

Proof: Consider any nonzero vector $v \in W_j$. Since π is a DSD, $v = \sum_{i \in I} v_i$ where $v_i \in V_i$ so we can divide v into two parts: $v = \sum_{V_i \subseteq W_j} v_i + \sum_{V_{i'} \not\subseteq W_j} v_{i'}$. Now $\sigma \preceq \pi$, so for each $v_{i'} \in V_{i'} \not\subseteq W_j$, there is a $W_{j'}$ such that $v_{i'} \in V_{i'} \subseteq W_{j'}$ so $\sum_{V_{i'} \not\subseteq W_j} v_{i'} \in \sum_{j' \neq j} W_{j'}$. But $\sum_{V_{i'} \not\subseteq W_j} v_{i'} = v - \sum_{V_i \subseteq W_j} v_i \in W_j$ and $W_j \cap \sum_{j' \neq j} W_{j'} = \{0\}$ since σ is a DSD. Thus $v - \sum_{V_i \subseteq W_j} v_i = 0$ so $v \in \oplus \{V_i \in \pi : V_i \subseteq W_j\}$. $\square$

Then $\sigma \preceq \pi$ implies $\pi \leftrightarrow \sigma$ and $\pi \vee \sigma = \pi$ as well as $\pi \wedge \sigma = \sigma$ as expected.

Proposition 18 *For any two DSDs π and σ, if they a common upper bound τ, i.e., $\pi, \sigma \preceq \tau$, then (i) $\pi \leftrightarrow \sigma$, and (ii) the join $\pi \vee \sigma$ is defined which is the least upper bound of π and σ.*

Proof: Since $\pi, \sigma \preceq \tau = \{X_k\}_{k \in K}$, then for each X_k, there is a $V_i \in \pi$ such that $X_k \subseteq V_i$ and there is a $W_j \in \sigma$ such that $X_k \subseteq W_j$ so $X_k \subseteq V_i \cap W_j$. Since the $\{X_k\}_{k \in K}$ span the space so must the nonzero $V_i \cap W_j$ so $\pi \leftrightarrow \sigma$ which proves (i) and makes $\pi \vee \sigma = \{V_i \cap W_j \neq \{0\}\}_{(i,j) \in I \times J}$ into a DSD. To prove (ii), as just shown, for any given X_k, there is a V_i and W_j such that $X_k \subseteq V_i \cap W_j$ so $\pi \vee \sigma$ is the least upper bound of π and σ in the refinement partial order. $\square$

Unlike set partitions on the same set, two DSDs π and σ need not have a common upper bound so $DSD(V)$ is not a join-semilattice.

Lemma 100 *Given a DSD $\pi = \{V_i\}_{i \in I}$, let $X = \oplus_{i \in I_X} V_i$ and $Y = \oplus_{i \in I_Y} V_i$ both be direct sums of some V_i's. If $X \cap Y$ is nonzero, then $X \cap Y = \oplus_{i \in I_X \cap I_Y} V_i$.*

Proof: For any nonzero $v \in X \cap Y$, there is a unique expression $v = \sum_{i \in I_X} v_{i,X}$ where $v_{i,X} \in V_i \subseteq X$ and a unique expression $v = \sum_{i \in I_Y} v_{i,Y}$ where $v_{i,Y} \in V_i \subseteq Y$. Since π is a DSD, there is also a unique expression $v = \sum_{i \in I} v_i$ so, for each nonzero v_i, $v_i = v_{i,X} = v_{i,Y} \in V_i \cap X \cap Y$. Thus for any such i, V_i is a common direct summand to X and Y, so $V_i \subseteq X \cap Y$. Thus every nonzero element $v \in X \cap Y$ is in a direct sum of V_i's for $V_i \subseteq X \cap Y$ and thus $X \cap Y$ is the direct sum of V_i that are common direct summands of X and Y. $\square$

Proposition 19 *The meet $\pi \wedge \sigma$ is the greatest lower bound of π and σ.*

Proof: If $\tau \preceq \pi, \sigma$ then each $X_k = \oplus\{V_i : V_i \subseteq X_k\} = \oplus\{W_j : W_j \subseteq X_k\}$. By the construction of $\pi \wedge \sigma$, there is a set partition $\{I_l\}_{l\in L}$ on I and a set partition $\{J_l\}_{l\in L}$ on J such that each subspace in the meet $\pi \wedge \sigma = \{Y_l\}$ satisfies: $Y_l = \oplus_{i\in I_l} V_i = \oplus_{j\in J_l} W_j$, and where no subsets of I smaller than I_l and subsets of J smaller than J_l have that property. Since each V_i is contained in some X_k, if $i \in I_l$, then $V_i \subseteq Y_l \cap X_k$. Since both Y_l and X_k are direct sums of some V_i, then by the Lemma the nonzero subspace $Y_l \cap X_k$ is also a direct sum of the common direct summand V_i's. Symmetrically, since the same Y_l and X_k are direct sums of some W_j's, then by the Lemma the nonzero subspace $Y_l \cap X_k$ is also a direct sum of the common direct summand W_j's. But since Y_l is the smallest direct sum of both V_i's and W_j's, $Y_l \cap X_k = Y_l$, i.e., $Y_l \subseteq X_k$, and thus $\pi \wedge \sigma$ is the greatest (in the refinement partial ordering) lower bound on π and σ. $\qquad\square$

Since the blob $\mathbf{0}$ is a lower bound for all DSDs, the meet of two DSDs always exists, which means that $DSD(V)$ is a meet semi-lattice.

The binary DSDs $\alpha = \{A_1, A_2\}$ are the atoms of the meet-semi-lattice $DSD(V)$, i.e., the DSDs so that there are no DSDs between them and the blob $\mathbf{0}$. A meet-semi-lattice is said to be *atomistic* if every non-blob element is a join of atoms and $DSD(V)$ is atomistic since any non-blob DSD $\pi = \{V_i\}_{i\in I}$ is clearly the join of the atoms $\{V_i, \oplus_{i'\in I, i'\neq i} V_{i'}\} \precsim \pi$.

4.6   The partition lattice determined by a maximal DSD

There is no maximum DSD, only maximal DSDs. Each maximal DSD in the partial ordering is a *discrete* or *completely decomposed* DSD of one-dimensional subspaces (or rays) of V (so the number of blocks is the dimension of V). Each basis set $U = \{u_1, ..., u_n\}$ determines a maximal DSD $\omega = \{[u_i]\}_{i=1}^n$ of the one-dimensional subspaces $[u_i]$ generated by the basis elements. A maximal ω determines the segment $[\mathbf{0}, \omega] = \{\pi | \pi \preceq \omega\}$ of $DSD(V)$ between 0 and ω. Fixing a basis set U so that $\omega = \{[u_i]\}_{i=1}^n$ then associates a set partition block B with every V_i of the vector space partition or DSD $\pi \precsim \omega$, namely the elements of the basis set U that generate V_i, so that the segment $[\mathbf{0}, \omega]$ with the induced operations of join and meet is isomorphic with the set partition lattice $\Pi(U)$. Or we could think of the one-dimensional subspaces as just points in a set ω so the segment is isomorphic to the partition lattice $\Pi(\omega)$ on the points ω.

$$\Pi(\omega) \cong \{\pi | \pi \preceq \omega\} = [\mathbf{0}, \omega] \subseteq DSD(V)$$

Thus $DSD(V)$ can be viewed as a set of overlapping partition lattices (or partition algebras when more operations are added) where a particular partition algebra is picked out by picking a maximal DSD. This relationship in the quantum logic of DSDs is analogous to the way in which a complete set of one-dimensional subspaces determines a Boolean algebra in the usual quantum logic of subspaces when viewed as a set of overlapping Boolean algebras or a "partial Boolean algebra" [55, p. 193]. Figure 4.2 illustrates the general 'shape' of $DSD(V)$ with the partition lattices determined by picking out a maximal DSD ω.

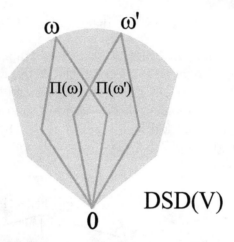

Figure 4.2: General shape of $DSD(V)$ with partition lattices determined by discrete DSDs.

For vector spaces over finite fields, e.g., $\mathbb{Z}_2^n$, the number of DSDs can be computed. Over a finite field with q elements, each discrete DSD determines $(q-1)^n$ bases since there are $q-1$ choices out of each one-dimensional subspace to be its basis element, but for $q = 2$, $(q-1)^n = 1$ so the number of discrete DSDs and the number of basis sets are the same in $\mathbb{Z}_2^n$. In $\mathbb{Z}_2^2$, there are only three two-element bases, i.e., three discrete DSDs, and no other DSDs except the indiscrete one $\mathbf{0}$. Thus fixing a basis ω determines the partition lattice $\Pi(\omega) \cong \wp(1)$ which we saw before was isomorphic to the two-element Boolean algebra $\wp(1)$. The vectors in a vector space over $\mathbb{Z}_2$ can be represented as subsets of a computational basis set $U = \{a, b\}$ where the addition of vectors is the symmetric difference $S + T = (S - T) \cup (T - S)$ of subsets. Thus for example $\{a\} + \{a, b\} = \{b\}$ where over $\mathbb{Z}_2$, the addition of elements is mod (2) so that $\{a\} + \{a\} = \emptyset$. For the sake of brevity in the figures, the zero vector (empty set) is left out of the subspaces in the illustrated DSDs and the elements in a subset are shorted to juxtaposition so that $\{a, b\}$ is written as $\{ab\}$. Thus the DSD $\{\{\emptyset, \{a\}\}, \{\emptyset, \{a, b\}\}\}$ is shorted to $\{\{a\}, \{ab\}\}$ in Figure 4.3.

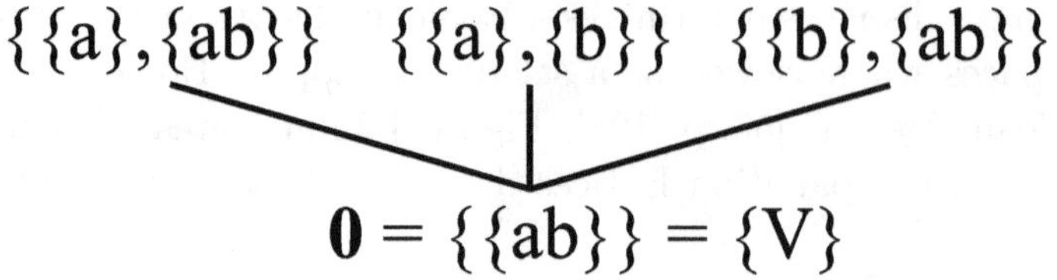

$$\{\{a\},\{ab\}\} \quad \{\{a\},\{b\}\} \quad \{\{b\},\{ab\}\}$$

$$\mathbf{0} = \{\{ab\}\} = \{V\}$$

Figure 4.3: The Hasse diagram for $DSD\left(\mathbb{Z}_2^2\right)$.

As n increases, the number of DSDs in $\mathbb{Z}_2^n$ increases rapidly, so for $n = 3$, there are 28 basis sets and also 28 DSDs between the 28 discrete DSDs and the one indiscrete one. Each discrete DSD has three blocks and those three blocks can be combined (i.e., direct summed) in $\binom{3}{2} = 3$ ways to form the three two-block DSDs between the three-block discrete DSD and the one-block blob $\mathbf{0}$. Figure 4.4 illustrates the fragment of $DSD\left(\mathbb{Z}_2^3\right)$ for two discrete DSDs, each with their three intermediate DSDs.

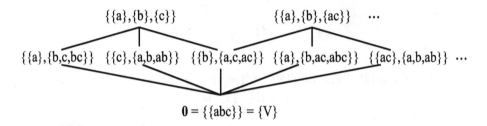

Figure 4.4: The part of $DSD\left(\mathbb{Z}_2^3\right)$ for two discrete DSDs, i.e., two different basis sets.

The fragment of $DSD\left(\mathbb{Z}_2^3\right)$ shown in Figure 4.4 is useful to illustrate the non-transitivity of compatibility. The DSD $\{\{b\}, \{a, c, ac\}\}$ and the DSD $\{\{c\}, \{a, b, ab\}\}$ to its left in Figure 4.4 have proto-join is $\{\{a\}, \{b\}, \{c\}\}$ which spans the whole space so they are compatible. To the right of $\{\{b\}, \{a, c, ac\}\}$ in Figure 4.4 is the DSD $\{\{a\}, \{b, ac, abc\}\}$ and their proto-join is $\{\{a\}, \{b\}, \{ac\}\}$ which also spans the space so they are also compatible. But for the two DSDs on the left and right of $\{\{b\}, \{a, c, ac\}\}$ in Figure 4.4, their proto-join is $\{\{a\}, \{b\}\}$ which does not span the space so they are not compatible.

Picking a basis set such as $\omega = \{\{a\}, \{b\}, \{c\}\}$ for $\mathbb{Z}_2^3$ then determines a segment $[\mathbf{0}, \omega]$ that is isomorphic to the set partition lattice on a three element set where each ray in the discrete DSD is treat as a point. Thus for this basis, the set partition lattice is illustrated in Figure 4.5 where a block $\{a, b\}$ in the set partition is abbreviated as ab.

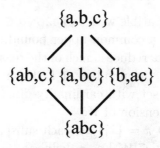

Partition lattice

Figure 4.5: Partition lattice isomorphic to segment $[\mathbf{0}, \omega]$ for
$$\omega = \{\{a\}, \{b\}, \{c\}\}.$$

If the basis set $\omega = \{\{ac\}, \{bc\}, \{abc\}\}$ had been chosen, then the segment $[0, \omega]$ would still be isomorphic to the set partition lattice on a three-element set where the three elements are denoted ac, bc, and abc, as illustrated in Figure 4.6.

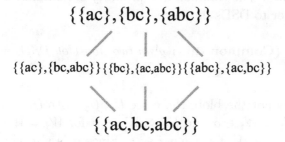

Figure 4.6: The partition lattice determined by another three-element basis set for $\mathbb{Z}_2^3$.

To fix notation, let an arbitrary maximal DSD be $\omega = \{U_z : z \in Z\}$ where each U_z is a one-dimensional subspace or ray and $|Z| = \dim(V)$. For any $\pi \in [\mathbf{0}, \omega]$, $\pi \preceq \omega$ so ω is (by definition) the maximum or *top* DSD in $[\mathbf{0}, \omega]$ and thus might be symbolized as the discrete DSD $\mathbf{1}_\omega = \omega$. Each subspace

$V_i \in \pi \preceq \omega$ has $V_i = \oplus \{U_z : U_z \subseteq V_i, z \in Z\}$ so $\mathbf{1}_\omega$ absorbs what it joins and is the unit element for meets within $[\mathbf{0}, \omega]$:

$$\pi \vee \mathbf{1}_\omega = \mathbf{1}_\omega \text{ and } \pi \wedge \mathbf{1}_\omega = \pi.$$

All the DSDs π and σ compatible with ω, i.e., $\pi, \sigma \in [\mathbf{0}, \omega]$, are compatible with each other since they have a common upper bound.

Fixing a maximal DSD ω reduces much of the reasoning in $[\mathbf{0}, \omega]$ to reasoning about sets. For instance, the number of DSDs in the segment $[\mathbf{0}, \omega]$ is the number of partitions on a set with cardinality $\dim(V)$, i.e., $\|[\mathbf{0}, \omega]\|$ is the Bell number [90] for n, the dimension of V.

Indeed, given any DSD $\pi = \{V_i\}_{i \in I}$, each subspace W_j of $\sigma \in [\mathbf{0}, \pi]$ determines a subset $C_j = \{V_i : V_i \subseteq W_j\}$ so σ defines a set partition $\sigma(\pi) = \{C_j\}_{j \in J}$ on π as a set so $W_j = \oplus C_j$ for $j \in J$. Thus the lower segment $[\mathbf{0}, \pi]$ is isomorphic to the set-based partition lattice (join and meet operations) on that set π, and, in particular, $[\mathbf{0}, \omega]$ is isomorphic to the lattice of set partitions on the set ω. As a partition lattice, $[\mathbf{0}, \pi]$, has the usual properties of partition lattices. Many theorems about set partitions can then be transferred over in an appropriate form to $[\mathbf{0}, \omega]$.

A *distinction* or *dit* of a DSD $\pi = \{V_i\}_{i \in I} \in [\mathbf{0}, \omega]$ for $\omega = \{U_z\}_{z \in Z}$ is a pair $(U_z, U_{z'})$ in distinct subspaces, i.e., $U_z \subseteq V_i$ and $U_{z'} \subseteq V_{i'}$ for some distinct $V_i, V_{i'} \in \pi$. In terms of eigenvectors and eigenvalues of a diagonalizable linear operator $F : V \to V$, a distinction or dits is a pair of eigenvectors of F with different eigenvalues. The common-dits property of non-blob set partitions [25, p. 106] carries over to DSDs in $[\mathbf{0}, \omega]$.

Proposition 20 (Common dits) *Any two non-blob DSDs $\pi, \sigma \in [\mathbf{0}, \omega]$ have a dit in common.*

Proof: Since π is not the blob, there are U_z, $U_{z'}$ with $U_z \subseteq V_i$ and $U_{z'} \subseteq V_{i'}$ for $V_i \neq V_{i'}$. If $U_z \subseteq W_j \in \sigma$ and $U_{z'} \subseteq W_{j'} \in \sigma$ for $W_j \neq W_{j'}$ we are finished so assume $U_z \oplus U_{z'} \subseteq W_j$ for some $j \in J$. Since σ is also not the blob, there is a $U_{z''}$ contained in some $W_{j''}$ where $W_{j''} \neq W_j$. Then $U_{z''}$ cannot be in the same subspace of π as U_z and $U_{z'}$ since those two are in different subspaces of π, so either $(U_z, U_{z''})$ or $(U_{z'}, U_{z''})$ is a dit common to π and σ. $\qquad\square$

For instance, in the segment $[\mathbf{0}, \omega]$ for $\omega = \{\{a\}, \{b\}, \{ac\}\}$ of Figure 4.4, there are three intermediate DSDs and $\binom{3}{2} = 3$ pairs of intermediate DSDs: $(\{a\}, \{b\})$ is a common dit of the DSDs $\{\{b\}, \{a, c, ac\}\}$ and $\{\{a\}, \{b, ac, abc\}\}$, and $(\{a\}, \{ac\})$ is a common dit of $\{\{a\}, \{b, ac, abc\}\}$ and $\{\{ac\}, \{a, b, ab\}\}$, while $(\{b\}, \{ac\})$ is a common dit of $\{\{b\}, \{a, c, ac\}\}$ and $\{\{ac\}, \{a, b, ab\}\}$.

As is appropriate to anything called a "logic," there should be an implication operation. Given the close connections between set partitions and vector-space partitions or DSDs given a maximal element ω, the DSD implication can be easily defined within $[\mathbf{0}, \omega]$.

Definition 101 *For $\sigma, \pi \in [\mathbf{0}, \omega]$, implication is:*

$$\sigma \Rightarrow_\omega \pi = \{U_z | U_z \subseteq V_i \text{ if } \exists V_i \in \pi \text{ and } W_j \in \sigma, V_i \subseteq W_j\}$$
$$\cup \{V_i | V_i \in \pi \text{ and there is no } W_j \in \sigma, V_i \subseteq W_j\}.$$

Since $V_i = \oplus \{U_z : U_z \subseteq V_i\}$, the implication is a DSD. For instance, for $\mathbb{Z}_2^4$ with $\omega = \{\{a\}, \{b\}, \{c\}, \{d\}\} = \mathbf{1}_\omega$, $\sigma = \{\{a\}, \{b, c, d, bc, bc, cd, bcd\}\}$ and $\pi = \{\{a\}, \{b\}, \{c, d, cd\}\}$, then $\sigma \Rightarrow_\omega \pi = \{\{a\}, \{b\}, \{c\}, \{d\}\} = \mathbf{1}_\omega$ since there is the inclusion of subspaces $\{c, d, cd\} \subseteq \{b, c, d, bc, bc, cd, bcd\}$. But the definition of the implication illustrates *contextuality* in the sense that it may differ depending on which segment is being considered. In the above example, $\omega' = \{\{a\}, \{b\}, \{c\}, \{cd\}\} = \mathbf{1}_{\omega'}$ is also a maximal element in $DSD\left(\mathbb{Z}_2^4\right)$ and both $\pi, \sigma \in [\mathbf{0}, \omega']$, and there is the same inclusion of subspaces so the subspace $\{c, d, cd\}$ is atomized to its one-dimensional components which are now $\{c\}$ and $\{cd\}$ so that:

$$\sigma \Rightarrow_{\omega'} \pi = \{\{a\}, \{b\}, \{c\}, \{cd\}\} = \mathbf{1}_{\omega'} \neq \mathbf{1}_\omega = \{\{a\}, \{b\}, \{c\}, \{d\}\} = \sigma \Rightarrow_\omega \pi.$$

In this case, both the contextualized implications were equal to $\mathbf{1}_\omega$ and $\mathbf{1}_{\omega'}$ which means in either case that $\sigma \precsim \pi$ (refinement is not a contextual notion).

Recall that for set partitions $\pi = \left\{f^{-1}(r)\right\}_{r \in f(u)}$ and $\sigma = \left\{g^{-1}(r)\right\}_{r \in g(U)}$, given by numerical attributes or random variables $f, g : U \to \mathbb{R}$, then $\sigma \Rightarrow \pi = \mathbf{1}_U$ iff $\sigma \precsim \pi$ which meant that f was a sufficient statistic for g. If the value of f was known for a experiment, then the value of g was determined. In the quantum context, F and G would be compatible observables determining DSDs $\pi = \{V_i\}_{i \in I}$ of the eigenspaces of F and $\sigma = \{W_j\}_{j \in J}$ of the eigenspaces of G, and ω as a maximal DSD of simultaneous eigenvectors for F and G. Then $\sigma \Rightarrow_\omega \pi = \mathbf{1}_\omega$, i.e., $\sigma \precsim \pi$, would mean that F was a sufficient observable for G. In any measurement using the measurement basis ω, the eigenvalue returned by an F-measurement determines the eigenvalue of the observable G. In general, $\sigma \Rightarrow_\omega \pi$ is the DSD in $[\mathbf{0}, \omega]$ that gives the extent to which the eigenvalue returned by an F-measurement determines the eigenvalue of G.

The one-dimensional subspaces U_z in the DSD $\sigma \Rightarrow \pi$ give the f eigenvalues, i.e., $U_z \subseteq V_i$, that determine the g eigenvalues. For instance if g had degenerate eigenvalues and $F_{\pi_1},...,F_{\pi_m}$ were observables with DSDs $\pi_1, ..., \pi_m$ also in $[\mathbf{0}, \omega]$

(and thus compatible), then $\sigma \Rightarrow \vee_{i=1}^{m} \pi_i = \mathbf{1}_\omega$ implies that the eigenvalues of $F_{\pi_1},...,F_{\pi_m}$ are sufficient to uniquely determine the eigenvalues of g. When $\vee_{i=1}^{m} \pi_i = \mathbf{1}_\omega$ as well, then the eigenvalues of $F_{\pi_1},...,F_{\pi_m}$ are sufficient to uniquely label the rays $U_z \in \omega$.

4.7   Exploiting duality in between the logics of subspaces and DSDs

The set partition operations (e.g., join, meet, and implication) on the partitions on a given universe set U can be represented as subset operations on certain subsets, i.e., ditsets, of $U \times U$. For a set partition $\pi = \{B_1, ..., B_m\}$ on U, a *distinction* or *dit of* π is an ordered pair $(u, u') \in U \times U$ of elements in distinct blocks of π. The ditset $\mathrm{dit}(\pi)$ of π is a binary relation on U (i.e., a subset of $U \times U$), and it is the complement in $U \times U$ of the equivalence relation associated with π. A partition relation on $U \times U$ is defined as the complement of an equivalence relation. The partition relations on $U \times U$ are in one-to-one correspondence with the partitions on U. Given a partition π on U, the ditset $\mathrm{dit}(\pi)$ is the corresponding partition relation, and given a partition relation, the equivalence classes in the complementary equivalence relation give the corresponding partition.

We have seen that the operations on the set partitions (join, meet, and implication) have corresponding operations on ditsets. Since $\sigma \preceq \pi$ iff $\mathrm{dit}(\sigma) \subseteq \mathrm{dit}(\pi)$, the partial order of refinement between partitions is just inclusion between ditsets. In this manner the partition algebra $\Pi(U)$ of partitions on U is represented as the algebra of the ditsets of $U \times U$.

With $\omega = \{U_z\}_{z \in Z}$ fixed and playing the role of U, the above construction can be transferred to vector spaces. The operations on DSDs in $\Pi(\omega)$ (the segment $[\mathbf{0}, \omega]$ endowed with the lattice and implication operations) can be represented as subspace operations on certain subspaces of the tensor product $V \otimes V$ that are direct sums of the subspaces in the maximal DSD $\omega \otimes \omega = \{U_z \otimes U_{z'} | (U_z, U_{z'}) \in \omega \times \omega\}$ of one-dimensional subspaces on $V \otimes V$. The easiest translation uses the fact that a DSDs $\pi = \{V_i\}_{i \in I} \in \Pi(\omega)$ defines a set partition $\pi(\omega) = \{B_i\}_{i \in I}$ on $\omega = \{U_z\}_{z \in Z}$ as a set where: $B_i = \{U_z | U_z \subseteq V_i\}$ and $V_i = \oplus B_i = \oplus \{U_z | U_z \subseteq V_i\}$ for $i \in I$. Then the *ditspace* $\mathrm{Dit}(\pi)$ defined by the DSD π is the following subspace of $V \otimes V$:

$$\mathrm{Dit}(\pi) = \oplus \{U_z \otimes U_{z'} | (U_z, U_{z'}) \in \mathrm{dit}(\pi(\omega))\}.$$

Note that by the Common-Dits Theorem, any two nonzero ditspaces, i.e., ditspaces for non-blob DSDs $\pi, \sigma \in \Pi(\omega)$, have a nonzero intersection. The operations on the ditspaces are those induced by the operations on the ditsets. For $\pi, \sigma \in \prod(\omega)$,

$$\text{Dit}\,(\pi \vee \sigma) = \oplus \{U_z \otimes U_{z'} \,|\, (U_z, U_{z'}) \in \text{dit}\,(\pi\,(\omega) \vee \sigma\,(\omega))\}$$
$$\text{Dit}\,(\pi \wedge \sigma) = \oplus \{U_z \otimes U_{z'} \,|\, (U_z, U_{z'}) \in \text{dit}\,(\pi\,(\omega) \wedge \sigma\,(\omega))\}$$
$$\text{Dit}\,(\sigma \Rightarrow \pi) = \oplus \{U_z \otimes U_{z'} \,|\, (U_z, U_{z'}) \in \text{dit}\,(\sigma\,(\omega) \Rightarrow \pi\,(\omega))\}.$$

The smallest ditspace is $\text{Dit}\,(\mathbf{0}) = \{0\}$ and the largest ditspace is $\text{Dit}\,(\mathbf{1}_\omega) = \oplus \{U_z \otimes U_{z'} | U_z \neq U_{z'}\}$, and the partial ordering is inclusion. Then the partition algebra of DSDs in $\prod(\omega)$ is represented by the algebra of the ditspaces of $V \otimes V$ for DSDs in $\prod(\omega)$.

Given the basic (category-theoretic) duality between subsets and partitions, this construction (using ditsets) to represent partition operations as subset operations–with the corresponding vector space version of the construction using ditspaces–has a dual construction to represent subset operations by partition operations. In the set case, instead of working with certain *subsets* (ditsets) of the *product* $U \times U$, the dual set construction works with certain *partitions* on the *coproduct* (disjoint union) $U \uplus U$. And for the vector space version, instead of working with subspaces (ditspaces) of the tensor product $V \otimes V$, the dual vector space construction works with DSDs on the coproduct or direct sum $V \oplus V^*$ (where V^* is a copy of V).

The set partition implication endows a rich structure on the partition algebra $\prod(U)$ of set partitions on U (always $|U| \geq 2$). For $\pi \in \prod(U)$, the π *-regular partitions* are the partitions of the form $\sigma \Rightarrow \pi$, which may be symbolized as $\overset{\pi}{\neg}\sigma$, for any $\sigma \in \prod(U)$. They are all in the segment $[\pi, \mathbf{1}_U]$ and they form a Boolean algebra, the *Boolean core* $\mathcal{B}_\pi$ of $[\pi, \mathbf{1}_U]$, under the partition operations of join, meet, and π-negation, where the π-negation of $\sigma \Rightarrow \pi = \overset{\pi}{\neg}\sigma$ is $(\sigma \Rightarrow \pi) \Rightarrow \pi = \overset{\pi}{\neg}\overset{\pi}{\neg}\sigma$. The dual construction uses this Boolean algebra based on partition operations.

We start with the set version of the dual construction and then go over the vector space version in more detail. Given a subset $S \subseteq U$, the *subset corelation* $\Delta\,(S)$ is the partition on the disjoint union $U \uplus U^*$ (U^* being a copy of U) whose blocks are the pairs $\{u, u^*\}$ for $u \in S$ and singletons $\{u\}$ and $\{u^*\}$ if $u \notin S$. Thus $\Delta\,(S)$ just encodes in a partition on the disjoint union which $u \in U$ are in the subset $S \subseteq U$. The subset corelations are partitions on the coproduct $U \uplus U$ defined by subsets of U, and they are dual to the relations that are subsets of the product $U \times U$. At the two extremes, $\Delta\,(U)$ is the bottom partition $\mathbf{0}_{U \uplus U^*}$ on $U \uplus U^*$ consisting of all pairs $\{u, u^*\}$ for $u \in U$, and $\Delta\,(\emptyset) = \mathbf{1}_{U \uplus U^*}$. The key lemma (see below) is that $(\Delta\,(S) \Rightarrow \Delta\,(U)) = \Delta\,(S^c)$ (analogous to $\sigma \Rightarrow \mathbf{0}_U = \neg\sigma$) so the $\Delta\,(U)$-negated partitions on $U \uplus U^*$ are the same as the complementary subset corelations $\Delta(S^c)$. Then it can be seen (proof below) that the Boolean core $\mathcal{B}_{\Delta(U)}$ of $[\Delta(U), 1_{U \uplus U^*}]$ is a Boolean algebra using the partition operations of join, meet, and $\Delta(U)$-negation that is isomorphic to the powerset BA $\wp(U)$ under the correspondence $\Delta(S) \Rightarrow \Delta(U) \mapsto S$ for $S \in \wp(U)$:

$$\mathcal{B}_{\Delta(U)} \cong \wp(U).$$

In that manner, the Boolean subset operations on subsets of U are represented by partition operations on certain partitions on $U \uplus U^*$[24, p. 320].

For the vector space version of the dual construction, note that given a maximal DSD $\omega = \{U_z\}_{z \in Z}$, there is the associated powerset BA $\wp(\omega)$ or $\wp(Z)$ depending on whether we take ω or Z as playing the role of U. Choosing the Z option, for each $S \subset \wp(Z)$, there is an associated subspace $A(S) = \oplus [U_z | z \subset S]$ and an associated projection operator $P_S : V \to V$ to that subspace. Each atomic DSD $\{A, A'\}$ in $\Pi(\omega)$ has the form $\{A(S), A(S^c)\}$ (where $S^c = Z - S$ is the complement in Z) with $V = A(Z)$ and $\{0\} = A(\emptyset)$. Thus there is an induced BA structure on the subspaces $\{A(S) | S \in \wp(Z)\}$ and on the projection operators $\{P_S | S \in \wp(Z)\}$ isomorphic to $\wp(Z)$. But how can this BA of certain subspaces of V be represented using the DSD operations of the logic of vector space partitions?

Let $V \oplus V^*$ be the direct sum (coproduct) of V with a copy V^* of itself. Given a maximal element $\omega = \{U_z\}_{z \in Z}$ of V, then the union with the copy $\omega^* = \{U_z^*\}_{z \in Z}$ forms a maximal element $\omega \cup \omega^*$ in the refinement ordering of DSDs in $DSD(V \oplus V^*)$ so we can work in the partition logic $\Pi(\omega \cup \omega^*)$.

Definition 102 *For $S \in \wp(Z)$ with the corresponding subspace $A(S)$, let $\Delta(A(S))$ or just $\Delta(S)$ be the DSD in $\Pi(\omega \cup \omega^*)$, called a subspace corelation, consisting of all the one-dimensional subspaces U_z and U_z^* for $z \notin S$, i.e., $U_z \nsubseteq A(S)$, and $U_z \oplus U_z^*$ for $z \in S$, i.e., $U_z \subseteq A(S)$.*

This is clearly just the subspace version of the subset corelation. Then $\Delta(Z)$ is the bottom $\mathbf{0}_{\omega \cup \omega^*}$ DSD consisting of all the subspaces $U_z \oplus U_z^*$ for $z \in Z$ and $\Delta(\emptyset) = 1_{\omega \cup \omega^*}$.

Lemma 103 $\Delta(S) \Rightarrow \Delta(Z) = \Delta(S^c)$.

Proof: For any $z \in S$, we have $U_z \oplus U_z^*$ in both $\Delta(S)$ and $\Delta(Z)$, so $U_z \oplus U_z^*$ is discretized in $\Delta(S) \Rightarrow \Delta(Z)$ into U_z and U_z^* separately. For any $z \in S^c$, $U_z \oplus U_z^*$ is only in $\Delta(Z)$ so it remains whole in $\Delta(S) \Rightarrow \Delta(Z)$ so that implication DSD is $\Delta(S^c)$. $\qquad\square$

Thus the $\Delta\left(Z\right)$-negated DSDs $\Delta\left(S\right)\Rightarrow\Delta\left(Z\right)$ are the subspace corelations in $\Pi\left(\omega\cup\omega^*\right)$. The Boolean core $\mathcal{B}_{\Delta(Z)}$ of the segment $[\Delta\left(Z\right),\omega\cup\omega^*]$ is a BA with the DSD operations of join, meet, implication, and $\Delta\left(Z\right)$-negation in $\Pi\left(\omega\cup\omega^*\right)$.

Proposition 21 $\mathcal{B}_{\Delta(Z)}\cong\wp\left(Z\right)$.

Proof: The isomorphism associates $\Delta\left(S\right)\Rightarrow\Delta\left(Z\right)\in\mathcal{B}_{\Delta(Z)}$ with $S\in\wp\left(Z\right)$. For $S,T\in\wp\left(Z\right)$, the union $S\cup T$ is associated with the join $\left(\Delta\left(S\right)\Rightarrow\Delta\left(Z\right)\right)\vee$ $\left(\Delta\left(T\right)\Rightarrow\Delta\left(Z\right)\right)=\Delta\left(S^c\right)\vee\Delta\left(T^c\right)=\Delta\left(S^c\cap T^c\right)=\Delta\left(\left(S\cup T\right)^c\right)=\Delta\left(S\cup T\right)$ $\Rightarrow\Delta\left(Z\right)$. The other Boolean operations of meet and $\Delta\left(Z\right)$-negation go in a similar manner. The null set $\emptyset\in\wp\left(Z\right)$ is associated with $\Delta\left(\emptyset\right)\Rightarrow\Delta\left(Z\right)=\Delta\left(\emptyset^c\right)=$ $\Delta\left(Z\right)$ which is the bottom of the BA $\mathcal{B}_{\Delta(Z)}$, and $Z\in\wp\left(Z\right)$ is associated with $\Delta\left(Z\right)\Rightarrow\Delta\left(Z\right)=\Delta\left(Z^c\right)=\Delta\left(\emptyset\right)=\mathbf{1}_{\omega\cup\omega^*}$ which is the top of $\mathcal{B}_{\Delta(Z)}$. If $S\subseteq T$ in $\wp\left(Z\right)$, then $T^c\subseteq S^c$ so $\Delta\left(S\right)\Rightarrow\Delta\left(Z\right)=\Delta\left(S^c\right)\preceq\Delta\left(T^c\right)=\Delta\left(T\right)\Rightarrow\Delta\left(Z\right)$ in the refinement ordering of $\Pi\left(\omega\cup\omega^*\right)$. $\qquad\square$

The treatment of *DSD operations* on V as *subspace* operations on $V\otimes V$, and the dual treatment of *subspace operations* on V as *DSD operations* on $V\oplus V^*$ exhibit the dual relationship between the two logics of DSDs and subspaces.

4.8   DSDs, CSCOs, and CSCDs

For a self-adjoint operator F on a Hilbert space V (or diagonalizable operator on any V), the projections P_{λ_i} can be constructed from the DSD $\pi=\{V_{\lambda_i}\}_{i\in I}$ of eigenspaces for the eigenvalues $\{\lambda_i\}_{i\in I}$, and then the operator can be reconstructed–given the eigenvalues–from the decomposition $F=\sum_{i\in I}\lambda_i P_{\lambda_i}$. A set partition $\pi=\{B,B',...\}$ on U can always be construed as the inverse-image partition of a numerical attribute $f:U\to\mathbb{R}$ by assigning different values to the different blocks in π. Thus the logic of set partitions is the logic of numerical attributes or random variables which is abstracted from the specific values and only reflects when the values were the same or different. In the same sense, the logic of DSDs of a vector space is the logic of diagonalizable operators on the space which is abstracted from the specific eigenvalues and only reflects when the eigenvalues were the same or different.

Given a state ψ and a self-adjoint operator $F:V\to V$ on a finite dimensional Hilbert space V, the operator determines the DSD $\pi=\{V_{\lambda_i}\}_{i\in I}$ of eigenspaces for the eigenvalues λ_i. The projective measurement operation uses the eigenspace DSD to decompose ψ into the unique parts given by the projections $P_{\lambda_i}\left(\psi\right)$ into the eigenspaces V_{λ_i}, where $P_{\lambda_i}\left(\psi\right)$ is the outcome of the projective measurement with probability $\Pr\left(\lambda_i|\psi\right)=\left\|P_{\lambda_i}\left(\psi\right)\right\|^2/\left\|\psi\right\|^2$.

The eigenspace DSD $\pi = \{V_{\lambda_i}\}_{i \in I}$ of F is refined by one or more maximal DSDs, i.e., $\pi = \{V_{\lambda_i}\}_{i \in I} \preceq \omega = \{U_z\}_{z \in Z}$. For each such ω, there is a set partition $\pi(\omega) = \{B_{\lambda_i}\}_{i \in I}$ on ω such that $V_{\lambda_i} = \oplus B_{\lambda_i}$. If some of the V_{λ_i} have dimension larger than one ("degeneracy"), then more measurements by commuting operators will be necessary to further decompose down to single dimensional eigenspace. If two self-adjoint operators commute, then their eigenspace DSDs are compatible. Given another self-adjoint operator $G : V \to V$ commuting with F, its eigenspace DSD $\sigma = \{W_{\mu_j}\}_{j \in J}$ (for eigenvalues μ_j of G) is compatible with $\pi = \{V_{\lambda_i}\}_{i \in I}$ and thus has a join DSD $\pi \vee \sigma$ in $DSD(V)$ which is also in $\Pi(\omega)$ for one or more maximal ω each representing an orthonormal basis of simultaneous eigenvectors. The combined measurement by the two commuting operators is just the single measurement using the join DSD $\pi \vee \sigma$.

Dirac's notion of a Complete Set of Commuting Operators (CSCO) $\{F_{\pi_l}\}_{l=1}^m$ [21] translates into the language of the quantum logic of DSDs as a *Complete Set of Compatible DSDs (CSCD)* $\{\pi_l\}_{l=1}^m$ whose join $\vee_{l=1}^m \pi_l$ *is a maximal DSD* $\omega = \mathbf{1}_\omega$ in $DSD(V)$ and thus is *the* maximum DSD $\mathbf{1}_\omega$ in $\Pi(\omega)$. As noted above, the eigenvalues of the observables F_{π_l} can then be used to uniquely label the $U_z \in \mathbf{1}_\omega = \omega$.

In partition logic on sets, a *valid formula*, i.e., a *partition tautology*, is a logical formula (using the partition operations of join, meet, and implication) so that when any partitions on the universe set U are substituted for the variables, the result is the discrete partition $\mathbf{1}_U$ on that set. Restated for DSDs, a *DSD tautology* in the partition logic $\Pi(\omega)$ for any maximal ω in $DSD(V)$ for any V is any formula (in the language of join, meet, and implication) so that no matter which DSDs of $\Pi(\omega)$ are substituted for the variables, the result is $\mathbf{1}_\omega$. For instance, *modus ponens* $\sigma \wedge (\sigma \Rightarrow_\omega \pi) \Rightarrow_\omega \pi$ is a DSD tautology in the partition logic $\Pi(\omega)$, so for any DSDs $\pi, \sigma \in \Pi(\omega)$, π is sufficient for $\sigma \wedge (\sigma \Rightarrow_\omega \pi)$. In the Boolean core $\mathcal{B}_\pi$ of $[\pi, \omega]$, the ordinary Boolean tautologies, like the law of excluded middle,

$$(\sigma \Rightarrow_\omega \pi) \vee ((\sigma \Rightarrow_\omega \pi) \Rightarrow_\omega \pi) = \neg_\omega^\pi \sigma \vee \neg_\omega^\pi \neg_\omega^\pi \sigma,$$

hold for any $\pi, \sigma \in \Pi(\omega)$, so they are DSD tautologies in the whole partition logic $\Pi(\omega)$, where that formula is the *weak* law of excluded middle for π-negation. Thus for any DSDs $\pi, \sigma \in \Pi(\omega)$, the DSDs $\sigma \Rightarrow_\omega \pi$ and $(\sigma \Rightarrow_\omega \pi) \Rightarrow_\omega \pi$ form a CSCD since their join is the discrete DSD $\mathbf{1}_\omega$. The law of excluded middle in $\mathcal{B}_\pi$ generalizes to the DSD tautology that is the disjunctive normal form decomposition of $\mathbf{1}_\omega$ for any number of variables. For instance, for any π, σ, and τ in $\Pi(\omega)$, we have the DSD tautology:

$$\left(\neg_\omega^\pi \neg_\omega^\pi \sigma \wedge \neg_\omega^\pi \neg_\omega^\pi \tau\right) \vee \left(\neg_\omega^\pi \neg_\omega^\pi \sigma \wedge \neg_\omega^\pi \tau\right) \vee \left(\neg_\omega^\pi \sigma \wedge \neg_\omega^\pi \neg_\omega^\pi \tau\right) \vee \left(\neg_\omega^\pi \sigma \wedge \neg_\omega^\pi \tau\right)$$

so those four disjuncts form a CSCD. Assigning distinct real numbers to the subspaces of the disjunct DSDs defines commuting Hermitian (or self-adjoint) operators that form one of Dirac's CSCOs.

4.9 Some concluding thoughts on the logic of partitions

It is early days in the development of the logic of set partitions and the logic of vector space partitions or direct-sum decompositions. In spite of the category-theoretic duality with subsets, partitions are considerably more complicated and that is reflected in the primal and dual structures in the partition algebras $\Pi(U)$. Compared to Boolean algebras, partition algebras are rather under-studied. In fact, there was over a century gap between the definition of the lattice operations of join and meet on partitions in the nineteenth century and the definition of the implication and other logical operations in the twenty-first century so that one could speak of "partition algebras" instead of just parti-tion lattices. Moreover, the study of "partition lattices" ([10]; [49]) using the upside-down lattice of equivalence relations did not promote the comparisons with the Boolean algebras or the Heyting and co-Heyting algebras.

The logic of subspaces specialized to Hilbert spaces, i.e., the usual quantum logic initiated by Birkhoff and Von Neumann [11], has been extensively, if not exhaustively, developed. The logic of the dual concept of direct-sum decomposi-tions was developed [28] using the sets-to-vector-spaces "Yoga of linearization". The development was in general terms without focusing exclusively on Hilbert spaces. Much of the quantum flavor can be brought out in the general case and even in the special case of vector spaces over $\mathbb{Z}_2$ [26].

The Boolean logic of subsets is usually treated as the logic of propositions by associating with each subset the proposition that a generic element of the uni-verse set is a member of the subset. Subset formulas that always evaluate to the universe set (i.e., subset tautologies) are then associated with the propositional formulas that always evaluate to True, i.e., truth-table tautologies. Boole [13] also developed the quantitative theory that measures subsets, i.e., probability theory. The quantum logic of subspaces is often characterized as the logic of the QM propositions that a state vector is in a subspace.

The logic of set partitions is the logic of numerical attributes or random variables that abstracts away from the numerical values and focuses on the inverse-image partition that retains the information as to whether the values were the same or different (i.e., the distinction between different blocks of the partition). If a proposition is to be associated with a partition, then it is the

proposition that the partition distinguishes a pair of different elements from
the universe set, i.e., that the pair of elements is a distinction of the partition
(two elements in different blocks of the partition). If a partition formula always
evaluates to the discrete partition then it is associated with the proposition
that the formula always distinguishes any pair of different elements, i.e., the
formula is a partition tautology. The quantitative theory that measures the
extent to which a partition distinguishes is the information theory of logical
entropy where the logical entropy of a partition is the probability that two
independent draws from the universe set give a distinction of the partition–just
as the probability of a subset is the probability that a single draw from the
universe set gives an element of the subset.[4] Linearization yields the logic of
direct-sum decompositions as the logic of diagonalizable linear operators (or
self-adjoint operators or observables in the QM case) that abstracts away from
the numerical values of the eigenvalues and that focuses on the DSD of the
eigenspaces that still reflects whether the eigenvalues were the same or different
(i.e., the distinction between the different eigenspaces). If a proposition is
to be associated with an observable in QM, then it is the proposition that the
observable distinguishes between its eigenvectors (by being in different blocks of
the DSD of eigenspaces). The quantum logical entropy of an observable applied
to a pure state is the probability that two independent projective measurements
of the pure state by that observable will yield a distinction of the observable's
DSD of eigenspaces, i.e., distinct eigenvalues.

Set concept	Linearization	Quantum version over Hilbert space
Boolean logic of subsets	Logic of subspaces	Orthomodular lattice of closed subspaces
Logic of partitions	Logic of DSDs	Quantum logic of DSDs

Table 4.2: Parallel development of subset and partition logics

The topic initially developed in this book is summarized in the bottom row of
Table 4.2.

One important application of partition logic is the quantitative or
information-theoretic versions of the logics of set partitions and vector-space
partitions which have been developed elsewhere ([27], [29], [32], [34]).

---

[4]Note the dual role of *elements* of a subsets and *distinctions* of a partition [31].

Chapter 5

Application 1: New Logical Foundation for Information Theory

5.1 Quantitative partition logic

After George Boole developed the logic of subsets, he developed the quantitative version which was the simplest Boole-Laplace notion of finite probability [13]. The probability of a subset $S \subseteq U$ is the number of its elements normalized: $\Pr\left(\frac{|S|}{|U|}\right)$. In view of the duality between elements of a subset and distinctions (dits) of a partition, the corresponding quantitative notion for a partition is automatic. But what sort of notion is it? Gian-Carlo Rota answered that question in his writings and MIT lectures. "The lattice of partitions plays for information the role that the Boolean algebra of subsets plays for size or probability." [66, p. 30] which he symbolized with the equivalence:

$$\frac{\text{Probability}}{\text{Subsets}} \approx \frac{\text{Information}}{\text{Partitions}}.$$

Hence the normalized number of dits in a partition will be a logical quantitative notion of information. Thus the *logical entropy* $h(\pi)$ of a partition $\pi = \{B_1, ..., B_m\}$ (with equiprobable points) on $U = \{u_1, ..., u_n\}$ is

$$h(\pi) = \frac{|\text{dit}(\pi)|}{|U \times U|} = \frac{|U \times U| - |\text{indit}(\pi)|}{|U \times U|} = 1 - \frac{\cup_j |B_j \times B_j|}{|U \times U|} = 1 - \sum_j \left(\frac{|B_j|}{|U|}\right)^2 =$$
$$1 - \sum_{j=1}^{m} \Pr(B_j)^2.$$

With point probabilities $p_1, ..., p_n$, the block probabilities are $\Pr(B_j) = \sum_{u_i \in B_j} p_i$, and the logical entropy of π is:

$$h\left(\pi\right) = 1 - \sum_{j=1}^{m} \Pr\left(B_j\right)^2.$$

The logical entropy is easy to interpret; in two independent draws from U, $\sum_{j=1}^{m} \Pr\left(B_j\right)^2$ is the probability of getting two elements in the same block of π, i.e., the indit-probability, so $h\left(\pi\right)$ is the two-draw probability of getting a distinction of π. In the equiprobable case, the logical entropy of the discrete partition is: $h\left(1_U\right) = 1 - \frac{1}{n}$, the probability that whatever was drawn the first time will not be drawn the second time. With the point probabilities $p = (p_1, ..., p_n)$, the logical entropy of the discrete partition is just taken as the logical entropy of the probability distribution:

$$h\left(p\right) = 1 - \sum_{i=1}^{n} p_i^2 = \sum_{i \neq k} p_i p_k$$

since squaring $1 = \sum_{i=1}^{n} p_i$ gives $1 - \sum_i p_i^2 = \sum_{i \neq k} p_i p_k$ (note that both $p_i p_k$ and $p_k p_i$ are included in the sum).

5.2   Logical entropy is a measure, a probability measure

The probability distribution p on U induces the product probability measure $p \times p$ on $U \times U$. The logical entropy $h\left(p\right)$ can then be written as $h\left(p\right) = \sum_{i \neq k} p_i p_k = p \times p\left(\text{dit}\left(1_U\right)\right)$, and in general,

$$h\left(\pi\right) = 1 - \sum_{(u_i, u_k) \in \text{indit}(\pi)} p_i p_k = \sum_{(u_i, u_k) \in \text{dit}(\pi)} p_i p_k = p \times p\left(\text{dit}\left(\pi\right)\right).$$

Thus logical entropy is a measure on $U \times U$ in the sense of measure theory [50].

The best-known notion of entropy is the Shannon entropy ([94]; [95]) which is defined directly in terms of a probability distribution p where the p_i's are the probabilities of the values of a random variable X:

$$H\left(X\right) = H(p) = \sum_{i=1}^{n} p_i \log_2\left(\frac{1}{p_i}\right).$$

For a partition, the Shannon entropy is defined in terms of the block probabilities:

$$H\left(\pi\right) = \sum_{j=1}^{m} \Pr\left(B_j\right) \log_2\left(\frac{1}{\Pr(B_j)}\right).$$

The logical and Shannon notions of entropy has been extensively compared [32] elsewhere so only a few points will be made here. Shannon defined the notions of simple, compound, conditional, and mutual Shannon entropy so that they satisfied the usual two variable Venn diagram relations as in Figure 5.1.

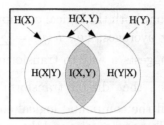

Figure 5.1: Venn diagram relations for compound Shannon entropies

But the problem is that the Shannon entropies are not defined as a measure on a set so the Venn diagrams, which usually illustrate a measure, are seen as just a mnemonic for the relationships between the compound notions of entropy.

Certain analogies between entropy and measure have been noted by various authors. These analogies provide a convenient mnemonic for the various relations between entropy, conditional entropy, joint entropy, and mutual information. It is interesting to speculate whether these analogies have a deeper foundation. It would seem to be quite significant if entropy did admit an interpretation as the measure of some set. [16, p. 112]

The logical entropy explains how the Shannon entropies can satisfy a Venn diagram relationship without being a measure. The compound notions of logical entropy satisfy the Venn diagram relations on the set $U \times U$ since the compound logical entropies were just the product measure applied to the appropriate sets in $U \times U$: $h(\pi \vee \sigma) = p \times p(\text{dit}(\pi \vee \sigma))$, $h(\pi | \sigma) = p \times p(\text{dit}(\pi) - \text{dit}(\sigma))$, $h(\sigma | \pi) = p \times p(\text{dit}(\sigma) - \text{dit}(\pi))$, and $m(\pi, \sigma) = p \times p(\text{dit}(\pi) \cap \text{dit}(\sigma))$ as illustrated in Figure 5.2.

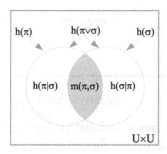

Figure 5.2: Venn diagram relations for compound logical entropies

The point to notice is that both the logical entropy and the Shannon entropy are defined as probability averages, i.e., $h\left(p\right) = \sum_i p_i \left(1 - p_i\right)$ and $H\left(p\right) = \sum_i p_i \log_2\left(\frac{1}{p_i}\right)$, so that suggests a way to transform the logical entropies into the Shannon entropies, i.e., the "dit-bit transform" $1 - p_i \rightsquigarrow \log_2\left(\frac{1}{p_i}\right)$ [32, p. 19]. In that manner, all the compound logical entropies transform into the corresponding compound Shannon entries and the non-linear but monotonic dit-bit transform preserves the Venn diagram relationships, e.g., $h\left(\pi\right) = h\left(\pi|\sigma\right) + m\left(\pi, \sigma\right)$. In that manner, logical entropy explains how the Shannon entropies can satisfy those relationships, e.g., $H\left(X\right) = H\left(X|Y\right) + I\left(X, Y\right)$, in spite of not being defined as a measure.

Shannon made a point about how the definitions gave the seemingly natural result that independent random variables had no mutual information: $I\left(X, Y\right) = 0$ for independent X and Y. But this got into trouble for three random variables which are pairwise independent but not mutually independent. Then the two-way overlaps have to be 0 but the three-way mutual information has to be non-zero–which means it has to be negative. A standard example is the throw of a pair of fair dice where $X = 1$ if first die gives an odd number, else 0, $Y = 1$ if the second die gives an odd number, else 0, and the $Z = 1$ if the sum $X + Y$ is odd, else 0. Then the Venn diagram for the Shannon entropies is as in Figure 5.3 [32, Figure 4.6].

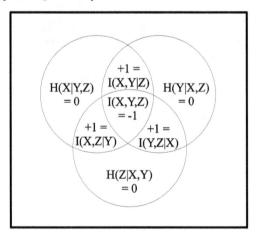

Figure 5.3: Example of negative Shannon mutual information

As it turns out, Shannon never defined mutual entropy $I(X, Y, Z)$ for three random variables even though the formula for it follows from the usual Venn diagram relationships, i.e., the inclusion-exclusion principle [93, Chapter 2]. There was no way to interpret negative mutual Shannon information.

> The set-function analogy might suggest to introduce further information quantities corresponding to arbitrary Boolean expressions of sets. E.g., the "information quantity" corresponding to $\mu(A \cap B \cap C) = \mu(A \cap B) - \mu(A \cap B - C)$ would be $I(X, Y) - I(X, Y|Z)$; this quantity has, however, no natural intuitive meaning. [20, pp. 52-3].

Another anomaly occurred with countable discrete probability distributions $p = (p_1, ...)$ where $\lim_{n \to \infty} \sum_{i=1}^{n} p_i = 1$. Since $\sum_{i=1}^{\infty} p_i = 1$, $\sum_{i=1}^{\infty} p_i^2$ is well-defined and thus so is $h(p)$. But for certain countable distributions, the Shannon entropy goes off to infinity ([115, p. 30]; [19, p. 48]).

Andrei Kolmogorov criticized Shannon entropy for starting with probabilities instead of basing information on something of a finite combinatorial character.

> Information theory must precede probability theory, and not be based on it. By the very essence of this discipline, the foundations of information theory have a finite combinatorial character. [64, p. 39]

Logical entropy satisfies the Kolmogorov desideratum since it is just the normalized count of the finite combinatorial object dit (π) which defines information-as-distinctions.

5.3 A brief history of the logical entropy formula

The logical entropy formula $h(p) = 1 - \sum_i p_i^2$ was motivated as the normalized count of the distinctions made by a partition when the probabilities are the block probabilities $p_B = \frac{|B|}{|U|}$ of a partition on a set U. The complementary measure $1 - h(p) = \sum_i p_i^2$ would be motivated as the normalized count of the identifications made by a partition thought of as an equivalence relation. Thus $1 - \sum_i p_i^2$, motivated by distinctions, is a measure of heterogeneity or diversity, while the complementary measure $\sum_i p_i^2$, motivated by identifications, is a measure of homogeneity or concentration. Historically, the formula can be found in either form depending on the particular context. The p_i's might be relative shares such as the relative share of organisms of the ith species in some population of organisms, and then the interpretation of p_i as a probability arises by considering the random choice of an organism from the population.

According to I. J. Good, the formula has a certain naturalness: "If $p_1, ..., p_t$ are the probabilities of t mutually exclusive and exhaustive events, any statistician of this century who wanted a measure of homogeneity would have take about two seconds to suggest $\sum p_i^2$ which I shall call ρ." [48, p. 561] The formula $1 - \sum p_i^2$ was used by Corrado Gini in 1912 ([44] reprinted in [45]) as a measure of diversity. But the strongest development of the formula (in the complementary form) in the early twentieth century was in cryptography. The American cryptologist, William F. Friedman, devoted a 1922 book ([42] reprinted as [43]) to the "index of coincidence" (i.e., $\sum p_i^2$) which, according to one author on the topic, "must be regarded as the most important single publication in cryptology." [59, p. 376] Two mathematicians, Solomon Kullback and Abraham Sinkov, who at one time worked as assistants to Friedman also wrote books on cryptology which used the index ([65] and [101]).

During World War II, Alan M. Turing worked for a time in the Government Code and Cypher School at the Bletchley Park facility in England. Probably unaware of the earlier work, Turing used $\rho = \sum p_i^2$ in his cryptoanalysis work and called it the *repeat rate* since it is the probability of a repeat in a pair of independent draws from a population with those probabilities (i.e., the identification probability $1 - h(p)$). Polish cryptoanalyists had independently used the repeat rate in their work on the Enigma [88].

After the war, Edward H. Simpson, a British statistician, proposed $\sum_{B \in \pi} p_B^2$ as a measure of species concentration (the opposite of diversity) where π is the partition of animals or plants according to species and where each animal or plant is considered as equiprobable. And Simpson gave the important interpretation of this homogeneity measure as "the probability that two individuals chosen at random and independently from the population will be found to belong to the same group." [100, p. 688] Hence $1 - \sum_{B \in \pi} p_B^2$ is the probability that a random ordered pair will belong to different groups. In the biodiversity literature [89], the formula is known as "Simpson's index of diversity" or sometimes, the "Gini-Simpson diversity index." However, Simpson along with I. J. Good worked at Bletchley during WWII, and, according to Good, "E. H. Simpson and I both obtained the notion [the repeat rate] from Turing." [47, p. 395] When Simpson published the index in 1948, he (again, according to Good) did not acknowledge Turing "fearing that to acknowledge him would be regarded as a breach of security." [48, p. 562]

In 1945, Albert O. Hirschman [53] suggested using $\sqrt{\sum p_i^2}$ as an index of trade concentration (where p_i is the relative share of trade in a certain commodity or with a certain partner). A few years later, Orris Herfindahl independently suggested using $\sum p_i^2$ as an index of industrial concentration (where p_i is the relative share of the ith firm in an industry). In the industrial economics literature, the index $H = \sum p_i^2$ is variously called the Hirschman-Herfindahl index, the HH index, or just the H index of concentration.

In view of the frequent and independent discovery and rediscovery of the formula $\rho = \sum p_i^2$ or its complement $1 - \sum p_i^2$ by Gini, Friedman, Turing, Hirschman, Herfindahl, and no doubt others, I. J. Good wisely advises that "it is unjust to associate ρ with any one person." [48, p. 562]

After Shannon's axiomatic introduction of his entropy [94], there was a proliferation of axiomatic entropies often with a variable parameter.[1] The formula $1 - \sum p_i^2$ for logical entropy (axiomatized in [84]) appeared as a special case for a specific parameter value in several cases. It appeared in 1967 as (half) of Havrda-Charvat's *structural* α *-entropy* [51]:

$$S(p_1, ..., p_n, ; a) = \frac{2^{a-1}}{2^{a-1}-1} \left(1 - \sum_i p_i^a\right)$$

for $a = 2$ and that special case was considered in 1969 by Vajda [107]. Patil and Taillie [81] defined the *diversity index of degree* β in 1982:

$$\Delta_\beta = \frac{1-\sum_i p_i^{\beta+1}}{\beta}$$

and Tsallis [105] independently gave the same formula as an entropy formula in 1988:

$$S_q(p_1, ..., p_n) = \frac{1-\sum_i p_i^q}{q-1}$$

where the logical entropy formula occurs as a special case ($\beta = 1$ or $q = 2$). While the generalized parametric entropies may be interesting as axiomatic exercises, our purpose is to emphasize the specific logical interpretation of the logical entropy formula (or its complement).

From the logical viewpoint, two elements from $U = \{u_1, ..., u_n\}$ are either identical or distinct so if $d_{ij} = 1 - \delta_{ij}$ was the "distance" between the ith and j^{th} elements, then the "logical distance" would be $d_{ij} = 1$ for $i \neq j$ and $d_{ii} = 0$. Since $1 = (p_1 + ... + p_n)(p_1 + ... + p_n) = \sum_i p_i^2 + \sum_{i\neq j} p_i p_j$, the logical entropy $h(p) = 1 - \sum_i p_i^2 = \sum_{i\neq j} p_i p_j$ is the probability that a pair of independently drawn elements are distinct, i.e., have logical distance of 1 between them. But

---

[1]There was no need for Shannon to present his entropy concept axiomatically since it was based on a standard concrete interpretation (expected number of binary partitions needed to distinguish a designated element) which could then be generalized. The axiomatic development encouraged the presentation of other "entropies" as if the axioms eliminated any need for an interpretation of the "entropy" concept.

one might generalize by allowing other distances $d_{ij} = d_{ji}$ for $i \neq j$ (but always $d_{ii} = 0$) so that $Q = \sum_{i \neq j} d_{ij} p_i p_j$ would be the average distance between a pair of independently drawn elements from U. The late C. R. (Calyampudi Radhakrishna) Rao introduced precisely this concept as *quadratic entropy* [83]. In many domains, it is quite reasonable to move beyond the bare-bones logical distance of $d_{ij} = 1 - \delta_{ij}$ so that Rao's notion of quadratic entropy is a useful and easily interpreted generalization of logical entropy.

5.4 Other points of comparison

Edwin Jaynes vigorously promoted a subfield in information theory, the Max-Entropy Program [58]. There is an old tradition that when there is no prior information about probabilities, then the best estimate is the Indifference Principle of assigning equal probabilities to the possible outcomes. That follows from maximizing either Shannon entropy or logical entropy in the absence of any constraints. But the Jaynes program was based on the idea that when the equiprobable distribution is ruled out by constraints, then it is best to choose the distribution that maximizes Shannon entropy. A standard problem used by Jaynes in lectures at Brandeis University is the Brandeis dice problem: What is the best probability distribution to assume for a die if the average of its throws is 4.5 instead of the average for a fair die of 3.5? [57, p. 47] The standard MaxEntropy method is to maximize the Shannon entropy (with natural logs) $H(p) = \sum_{i=1}^{6} p_i \ln\left(\frac{1}{p_i}\right)$ subject to the constraint $\sum_{i=1}^{6} p_i i = 4.5$. Using the method of Lagrange multipliers with that constraint, then the answers using the transcendental natural log function is to four decimal places:

$$p_{MaxH} = (0.0543, 0.0788, 0.1142, 0.1654, 0.2398, 0.3475).$$

If the logical entropy $h(p) = 1 - \sum_{i=1}^{6} p_i^2$ was maximized instead with the same constraint, the answers are rational numbers:

$$p_{Maxh} = \frac{1}{210}(5, 17, 29, 41, 53, 65)$$
$$= (0.0238, 0.0810, 0.1381, 0.1952, 0.2524, 0.3095).$$

The answer from maximizing $h(p)$ is different from maximizing $H(p)$ and that is generally true.

Hence the question arises of which maximization is the "best" prior probability when the equiprobable distribution is not available. Since probability distributions $p = (p_1, ..., p_n)$ are all in the non-negative orthant of $\mathbb{R}^n$, one plausible answer is that the best answer is the one closest to the equiprobable distribution using the standard notion of Euclidean distance $\sqrt{\sum_{i=1}^{n} \left(p_i - \frac{1}{n}\right)^2}$. Since minimizing the Euclidean distance gives the same answer as minimizing its square, we have:

$$\sum_{i=1}^{n} \left(p_i - \tfrac{1}{n}\right)^2 = \sum_i \left(p_i^2 - 2p_i\tfrac{1}{n} + \tfrac{1}{n^2}\right) = \sum_i p_i^2 - \tfrac{2}{n} + \tfrac{n}{n^2}$$
$$= \sum_i p_i^2 - \tfrac{1}{n} = \left(1 - \tfrac{1}{n}\right) - \left(1 - \sum_i p_i^2\right) = \left(1 - \tfrac{1}{n}\right) - h\left(p\right).$$

Hence minimizing the distance to the equiprobable distribution is the same as maximizing the logical entropy.

The equiprobable distribution has the minimal variance of 0 so another selection criterion would be to find the feasible probability distribution with the minimal variance. Taking the p_i's as the values of a random variable with equiprobable values, the variance of p is:

$$Var\left(p\right) = \sum_i \tfrac{1}{n}p_i^2 - \left(\sum \tfrac{1}{n}p_i\right)^2 = \tfrac{1}{n}\left[\sum_i p_i^2 - \tfrac{1}{n}\right] = \tfrac{1}{n}\left[\left(1 - \tfrac{1}{n}\right) - h\left(p\right)\right]$$

so minimizing the variance $Var\left(p\right)$ is the same as maximizing the logical entropy. When unconstrained, the maximum value of logical entropy is the equiprobable distribution where $h\left(p\right) = 1 - \frac{1}{n}$ (the probability the second draw is not the same as the first) which gives the variance of 0 for that equiprobable distribution.

The only alternative to maximizing Shannon entropy considered by Jaynes [58, pp. 345-6] was minimizing $\sum_i p_i^2$ which is the same as maximizing logical entropy. His criticism was that sometimes the minimizing of $\sum_i p_i^2$ by the Lagrange multiplier method would give negative probabilities. For instance if the Brandeis dice problem has the average of 5 instead of 4.5, then maximizing $h\left(p\right)$ by that Lagrange multiplier method would give negative probabilities. But $h\left(p\right)$ is a quadratic maximand so the proper optimization method in the Brandeis dice problem would be quadratic programming ([9]; [61]) which includes non-negativity constraints on the variables. Indeed, for the Brandeis dice problem with the required mean of 5, the two solutions are:

$$p_{MaxH} = (0.0205, 0.0385, 0.0723, 0.1357, 0.2548, 0.4781) \text{ and}$$
$$p_{Maxh} = \tfrac{1}{10}(0, 0, 1, 2, 3, 4),$$

where the (rational) logical entropy solution p_{Maxh} is closer to $\left(\frac{1}{6}, ..., \frac{1}{6}\right)$ and has lower variance than p_{MaxH}.

The logical entropy $h\left(X\right) = \sum_{i,k} p_i p_k \left(1 - \delta_{ik}\right)$ is a special case [84] of C. R. Rao's notion of quadratic entropy $Q\left(X\right) = \sum_{i,k} p_i p_k d\left(x_i, x_k\right)$ ([83]; [85])

where $d(x_i, x_k)$ is a non-negative symmetric notion of the distance between x_i and x_k. For the logical entropy, $1 - \delta_{ik}$ is the logical distance function where $d(x_i, x_k) = 1$ if $x_i \neq x_k$ and $d(x_i, x_k) = 0$ if $x_i = x_k$. The Shannon entropy is often described as a measure of uncertainty. However, there is already a measure of uncertainty in statistics, namely the variance $Var(X) = E(X^2) - E(X)^2$. In the spirit of logical entropy, the quadratic entropy measure of uncertainty would be the average distance or distance squared between the values of the random variable: $\sum_{i,k} p_i p_k (x_i - x_k)^2$. But there are two choices for the sum: all i, k which would count $p_i p_k (x_i - x_k)^2$ twice or just for $i < k$ which would only count it once. Then we have:

$$\sum_{i<k} p_i p_k (x_i - x_k)^2 = Var(X)$$

and, of course, $2Var(X)$ [63, p. 42] if counted twice.

This result can be extended to the covariance $Cov(X,Y) = E(XY) - E(X)E(Y)$ where $p_{ij} = \Pr(x_i, y_j)$ is the joint probability distribution for x_i with $i = 1, ..., n$ and y_j with $j = 1, ..., m$. Then the formula in the spirit of logical entropy[2] is:

$$\sum_{(i,j)\neq(i',j')} p_{ij} p_{i'j'} (x_i - x_{i'}) (y_j - y_{j'}) = 2Cov(X,Y)$$

and with the lexicographic ordering on the indices [32, p. 52];

$$\sum_{(i,j)<(i',j')} p_{ij} p_{i'j'} (x_i - x_{i'}) (y_j - y_{j'}) = Cov(X,Y).$$

The Shannon entropy $H_e(p) = \sum_i p_i \ln\left(\frac{1}{p_i}\right)$ is often associated with (and sometimes identified with) the Boltzmann entropy $S = \ln\left(\frac{n!}{n_1!...n_m!}\right)$. The Boltzmann entropy is not very analytically tractable so it is advantageous to use a numerical approximation. The Stirling series is an infinite series for $\ln(n!)$ but the first two terms in the series give the approximation: $\ln(n!) \approx n \ln(n) - n$. Several writers have noted that there are better approximations ([3, p. 533]; [75, p. 2]) but they all use the two-term Stirling approximation since when it is used, then the Boltzmann entropy is approximated by the Shannon entropy. Unlike the other approximations, the Shannon entropy approximation has very nice analytical properties, e.g., the partition function in statistical mechanics and thermodynamics. This two-term Stirling approximation has become so

---

[2]The formula is not a special case of quadratic entropy since $(x_i - x_{i'})(y_j - y_{j'})$ can be negative.

standard that some writers talk about the "Boltzmann-Shannon entropy" as if the notions were conceptually the same.

The question is the role of logical entropy in information theory. Shannon never called it "information theory"; it was the theory of communication. He even complained about the "information bandwagon" [96] that was the making of other writers and science popularizers. Shannon also noted that "no concept of information itself was defined" [97, p. 458] in communications theory but that his entropy was a quantification of information. That is correct since Shannon entropy is a monotonic transform (the dit-bit transform) of logical entropy.

In contrast, logical entropy is the measure of the number of distinctions in partitions, where the logic of partitions is the dual of the logic of subsets, and both notions are at the same level of mathematical fundamentality. Thus logical entropy does provide a definition of information, namely information as distinctions (or distinguishability or differences depending on the context). As Charles Bennett, one of the founders of quantum information theory, put it: "So information really is a very useful abstraction. It is the notion of distinguishability abstracted away from what we are distinguishing, or from the carrier of information...." [8, P. 155] The notion of logical entropy extends naturally to quantum logical entropy, e.g., $h(\rho) = 1 - \text{tr}[\rho^2]$ ([32]; [104]).

From the mathematical viewpoint, logical entropy is a measure in the sense of measure theory while Shannon entropy is not defined as a measure on a set. Moreover, the Shannon entropy exhibits several anomalies that do not afflict logical entropy. In the MaxEntropy application, logical entropy exhibits superior properties. The Shannon entropy exhibits its superior properties in the theory of communications and coding–as Shannon emphasized–and historically as an analytically tractable numerical approximation to Boltzmann entropy.

Our conclusion is that logical entropy, as the information quantification of partitions in Rota's equivalence $\frac{\text{Probability}}{\text{Subsets}} \approx \frac{\text{Information}}{\text{Partitions}}$, provides a "new foundation" [32], a *logical* foundation, for information theory. The compound formulas for Shannon entropy can all be derived as the dit-bit transform of the corresponding logical entropy formulas–which is ultimately due to the fact that Shannon entropy can also be defined as the minimum average number of binary *distinctions* (bits or yes-no questions) necessary to *distinguish* all the messages.

This new logical foundation for information theory is the first important application of the logic of partitions. Modern quantum theory was consolidated almost a century ago in the mid-1920's. Yet over the ensuing century, it has not been established how to physically interpret the theory. As David Mermin put it: "New interpretations appear every year. None ever disappear." [78, p.

8] The second major application of partitions is to provide a new mathematical basis for an interpretation of quantum mechanics that is not entirely new and which might be called the Literal [99, p. 6] or the Objective Indefiniteness Interpretation of Quantum Mechanics.

Chapter 6

Application 2: Interpreting Quantum Mechanics

6.1 Introduction: Quantum versus classical reality

The mathematics of quantum mechanics is quite distinctive and different from that of classical mechanics. Hence a natural strategy to understand and interpret QM is to "follow the math" back to its source ([35], [36], [37]). Our argument is that the source is the mathematics of set partitions; the math of QM is the Hilbert space extension of the math of partitions. Put the other way around, the math of partitions can be seen as a 'skeletionized' version of the math of QM–particularly when partition math is formulated in a vector space over $\mathbb{Z}_2$ [26].

Partitions are the logical concept to represent distinctions (dits) and indistinctions (indits), definiteness and indefiniteness, distinguishability and indistinguishability, inequivalence and equivalence, or difference and identity. One of the characteristic features of QM, now widely recognized in textbooks and the quantum literature, is that prior to a measurement, a superposition state does not have a definite value of the observable being measured. It is indefinite and the indefiniteness is objective in nature, i.e., it is not just that the observer is subjectively indefinite as to the value of the observable.

> From these two basic ideas alone – indefiniteness and the superposition principle – it should be clear already that quantum mechanics conflicts sharply with common sense. If the quantum state of a system is a complete description of the system, then a quantity that has an indefinite value in that quantum state is objectively in-

definite; its value is not merely unknown by the scientist who seeks to describe the system. [98, p. 47]

These few hints already tell us that the math of partitions should be related to the math of QM.

Furthermore, our classical intuitions (realized in classical physics) are that reality is fully definite–one might say "definite all the way down." That is the idea behind Leibniz's Principle of Identity of Indistinguishables (PII) [2, Fourth letter, 22] and Kant's Principle of Complete Determination.

> Every thing, however, as to its possibility, further stands under the principle of thoroughgoing determination; according to which, among all possible predicates of things, insofar as they are compared with their opposites, one must apply to it. [60, B600]

If reality is definite "all the way down," then for any two distinct entities, by digging deeper there is always a distinguishing predicate; otherwise the entities must be identical. The PII is false in QM.[1] Leibniz also had the Principle of Continuity that was expressed by "*Natura non facit saltus*" (Nature does not make jumps) [70, Bk. IV, chap. xvi] and his Principle of Sufficient Reason was expressed as the statement "that nothing happens without a reason why it should be so rather than otherwise" [2, Second letter, 7]. Both of these are also false in QM; continuity is violated by the infamous quantum jumps and sufficient reason is violated by the objective nature of quantum probabilities.

> Furthermore, since the outcome of a measurement of an objectively indefinite quantity is not determined by the quantum state, and yet the quantum state is the complete bearer of information about the system, the outcome is strictly a matter of objective chance – not just a matter of chance in the sense of unpredictability by the scientist. Finally, the probability of each possible outcome of the measurement is an objective probability. Classical physics did not conflict with common sense in these fundamental ways. [98, p. 47]

---

[1] If two bosons with the same intrinsic characteristics (mass, charge,...), which we name A and B, are in a pair state $|A\rangle |B\rangle$, then there is no way to mark or distinguish them from the permuted state $|B\rangle |A\rangle$ so their actual physical state is better represented as $|A\rangle |B\rangle + |B\rangle |A\rangle$ which is unchanged under permutation.

6.2 Quantum reality is indefinite world, not wave world

It seems the main roadblock to understanding how to interpret quantum mechanics over the last century has been the inclination to assign an ontological status to computational artifacts such as the wave function $\psi(x)$. The math of QM needs to use vector spaces over the algebraically complete extension $\mathbb{C}$ of the real field $\mathbb{R}$ so that (among other reasons) all the real-valued observables (Hermitian operators) will have a complete set of eigenvectors [110, 67, fn. 7]. The complex numbers $\mathbb{C}$ is the natural mathematics to describe waves, e.g., a complex number in the polar representation has an amplitude and phase. The addition of vectors in a vector space over $\mathbb{C}$ can always be interpreted in terms of the interference of waves [39, Chap. 29.5], so the wave-math is not wrong but is not something required to represent quantum reality. The wave-math comes free-of-charge in a vector space over $\mathbb{C}$. That addition-of-waves interpretation of the superposition of quantum states, illustrated in Figure 6.1, is not the way to interpret the key concept of superposition in quantum reality.

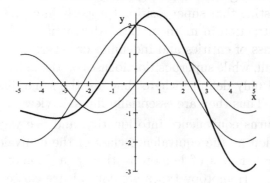

Figure 6.1: The addition-of-waves interpretation of superposition

Given a basis for a Hilbert space[2], say, an orthonormal (ON) basis set of eigenstates of an observable, a superposition is some linear combination of those basis eigenstates. What is the alternative way to interpret a superposition state? A superposition state should be interpreted as a state that is indefinite (with various amplitudes) on where the eigenstates differ and is definite only on where they are the same. Figure 6.2 gives some suggestive imagery using the 'superposition' of two isosceles triangles. The superposition is indefinite on

---

[2]Since our purpose is conceptual clarity, not mathematical generality, we will stick to the finite dimensional case for Hilbert spaces and finite universe sets for partitions.

where the triangles differ, the labeling of the equal sides B, C and equal angles b, c, but is definite on where they are the same, the angle a and opposing side A. In particular, it is not the "double exposure" picture where the triangle has double-labeled sides and angles where they differ.[3]

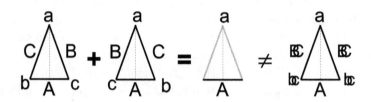

Figure 6.2: Indefiniteness interpretation of superposition (as opposed to the wave interpretation)

Since we are dealing with basic logical notions (indefiniteness and definiteness),[4] it is interesting that superposition (properly interpreted) is the flip-side of a notion of abstraction in mathematics. Abstraction is definite on the commonalities in a class of entities and indefinite on where they differ, like seeing the glass as half-full, while superposition focuses on the indefiniteness where the entities differ and only definite where they are the same, like seeing the glass as half-empty.[33] Thus they are essentially flip-side views of one basic notion.

Abstraction turns equivalence into identity, and one way to do that is to take the abstraction as the equivalence class of the equivalence entities. For instance, in older versions of homotopy theory, a "homotopy type" was an equivalence class. "Homotopy types are equivalence classes of spaces" under the equivalence relation "of deformation or homotopy." [7, p. 4] But in a recent treatment [106], the equivalence class of, say, unit-interval-coordinatized paths going once around the ring clockwise is abstracted to an object having those common characteristics (Figure 6.3) but is not the equivalence class or one of the paths in the equivalence class.

---

[3]That incorrect interpretation is common in the popular science literature which, for instance, describes a particle as going through both slits in the double-slit experiment. That is inspired by the image of a wave hitting the two slits at the same time as in classical classroom demonstrations of water wave interference.

[4]Unfortunately, the philosophical literature on vagueness (e.g., sorites arguments) seems to produced little of interest in the old problem of interpreting quanum mechanics as noted by Peter Lewis [71].

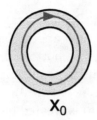

X_0

Figure 6.3: Abstract notion of a path once clockwise around a ring

This philosophical connection between superposition as focusing on indefiniteness-on-differences and abstraction as focusing on definiteness-on-commonalities is not available when thinking of superposition as wave-addition.

6.3 The two dual lattices tell different 'creation stories'

The two dual logics provide a modern model for the old duality of matter or *substance*[5] versus *form* (as in in-form-ation) [1]. If we consider moving up the two lattices from the bottom to the top, we see that substance increases in the subset lattice while form stays constant (fully distinct), while the reverse happens in the partition lattice.

For each lattice where $U = \{a, b, c\}$, start at the bottom and move towards the top in Figure 6.4.

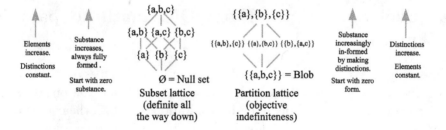

Figure 6.4: Moving up the subset and partition lattices.

---

[5]Heisenberg identifies "substance" with energy.

Energy is in fact the substance from which all elementary particles, all atoms and therefore all things are made, and energy is that which moves. Energy is a substance, since its total amount does not change, and the elementary particles can actually be made from this substance as is seen in many experiments on the creation of elementary particles. [52, p. 63]

There is no substance or "its" (elements) at the bottom $\emptyset$ of the subset lattice, and one moves up, new "its" appear but always fully formed until one reaches the full universe U. At the bottom $\mathbf{0}_U$ of the partition lattice, there are no dits (distinctions) and all the substance already appears but with no form (i.e., no dits in $\mathbf{0}_U$ just as no its in $\emptyset$). As one moves up the lattice of partitions, form (or in-form-ation) is created by making new dits or distinctions until reaching the partition $\mathbf{1}_U$ that distinguishes all elements ("its from dits")– while all the substance is present in each partition.

The progress from bottom to top of the two lattices could also be described as two creation stories.

- *Subset creation story*: "In the Beginning was the Void", and then elements are created, fully propertied and distinguished from one another, until finally reaching all the elements of the universe set U.

- *Partition creation story*: "In the Beginning was the Blob", which is an undifferentiated perfectly symmetrical "substance," and then there is a "Big Bang" where the substance is being objectively in-formed by the making of distinctions (e.g., breaking symmetries) until the result is finally the singletons which designate the elements of the universe U.

The partition creation story, can be viewed as a bare-bones description of the Big Bang theory of creation–constant substance (energy) in-formed by symmetry breaking. [80]

6.4 Skeletonization: From QM math to partition math

Skeletonizing is a technique to convert a quantum state such as $\alpha_i \left|u_i\right\rangle + \alpha_j \left|u_j\right\rangle + \alpha_k \left|u_k\right\rangle$ into its skeletonized set version $\{u_i, u_j, u_k\}$, that is, throw out the scalars $\alpha_i, \alpha_j, \alpha_k \in \mathbb{C}$, the ket-symbols $\left|\right\rangle$, and the addition to obtain the corresponding set as a block in a partition. For instance, in the four-dimensional Hilbert space $\mathbb{C}^4$ with an orthonormal basis set $\left|a\right\rangle, \left|b\right\rangle, \left|c\right\rangle, \left|d\right\rangle$, all the vectors would skeletonize to blocks over $U = \{a, b, c, d\}$. A superposition such as the pure state $\alpha_a \left|a\right\rangle + \alpha_b \left|b\right\rangle + \alpha_c \left|c\right\rangle + \alpha_d \left|d\right\rangle$ would skeletonize to the indiscrete partition $\mathbf{0}_U = \{\{a, b, c, d\}\}$–where we might use the shorthand of replacing the innermost curly brackets with juxtaposition so that $\mathbf{0}_U = \{abcd\}$. Then probabilistic mixed (orthogonal) states like $\alpha_a \left|a\right\rangle + \alpha_d \left|d\right\rangle$ and $\alpha_b \left|b\right\rangle + \alpha_c \left|c\right\rangle$ with probabilities that sum to one would skeletonize to the partition $\{\{a, d\}, \{b, c\}\}$ or, in shorthand, $\{ad, bc\}$. In this manner, the pure and certain mixed states in Hilbert space can be skeletonized to set partitions in the partition lattice $\Pi(U)$ of partitions on an orthonormal basis set for the Hilbert space as shown in Figure 6.5. Since not all the states in Hilbert space when represented in a certain ON basis U will skeletonize to a partition, this "skeletal model" is only meant as an aid to intuition to represent certain Hilbert space operations in terms of partition math. In the business of visualizing quantum reality, a super-simplified image is perhaps better that none at all.

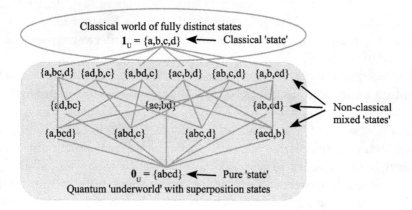

Figure 6.5: Skeletonized version of classical and quantum worlds

The non-classical mixed or pure states contain non-singleton blocks representing the basic non-classical notion of superposition. The classical world is represented by the completely mixed state of singletons, i.e., by the discrete partition $\mathbf{1}_U$. For instance, in the six-dimensional case, the discrete partition would represent "the statistical mixture describing the state of a classical dice before the outcome of the throw" [4, p. 176]. One of the older characterizations of classical reality was Leibniz's Principle of Identity of Indistinguishables. In a partition, the elements that are indistinguishable (by that partition) are the elements in the same block so they form an indistinction. The discrete partition satisfies the partition logic version of Leibniz's Principle.

For any $u, u' \in U$, if $(u, u') \in \text{indit}(\mathbf{1}_U)$, then $u = u'$
Partition logic Principle of Identity of Indistinguishables.

No other partition satisfies that notion of classicality since it would involve a non-singleton block.

This general picture of a quantum world involving objectively indefinite superpositions is not new. Any quantum theorist who recognizes that a superposition of states in the observable basis (different eigenvalues) does not have a definite value prior to measurement has already recognized that quantum reality has objective indefiniteness. Heisenberg's Indeterminacy Principle also makes the basic point that there is a necessary indefiniteness involved in conjugate observables; the more definite the value of one observable, the more indefiniteness in the value of a conjugate observable.

There is a host of quantum theorists that have extended that insight into recognizing a whole quantum world of indeterminacy, labelled "potentialities" or "latencies." Those theorists include Werner Heisenberg [52], Abner Shimony [99], Henry Margenau [76], R. I. G. Hughes [55], Ruth Kastner [62], and others. Kastner even likens the quantum underworld to the underwater part of an iceberg with only the classical world above the water–a picture akin to the Figure 5.5 skeletonized image of quantum states. One language used is to characterize the quantum underworld as a form of reality that is still a "potentiality" while the classical states are actualities. All these quantum thinkers agree that the quantum underworld of "potentialities" or "latencies" is characterized by objective indefiniteness. Hence this use of two modalities (real potentialities versus actualities) is unnecessary in the objective indefiniteness representation; it is only the difference between objectively indefinite and fully definite realities. Rather than introducing a new interpretation, the partition math approach only adds corroboration to those, like Heisenberg and Shimony, who have already recognized a quantum world characterized by objective indefiniteness–the indefinite world.

6.5   Partitions and quantum states

The application of partition math to quantum mechanics is based on the thesis that the distinctive math of QM (not the physics which is brought over from classical physics by quantization) is the Hilbert space version of set level math of partitions. There three distinctive concepts in the math of QM that need to be accounted for;

1. the notion of a quantum state,

2. the notion of a quantum observable, and

3. the notion of measurement (always projective) or state reduction (to use a less anthropomorphic term).

The notion of a quantum state is usually described in terms of the wave function which we have argued is ontologically misleading to understand quantum reality. Is there an equivalent mathematical tool that can be used instead of the wave function and which directly represents the objective indefiniteness? Yes, it is the density matrix [109].

Our strategy is to first reformulate a set partition $\pi = \{B_1, ..., B_m\}$ on $U = \{u_1, ..., u_n\}$ with (assumed all positive) point probabilities $p_1, ..., p_n$ as a density matrix $\rho(\pi)$ and then to show how the general case of a density matrix ρ in QM is the Hilbert space version of the partition version $\rho(\pi)$.

For each block B_j in π, we define the column vector $|b_j\rangle$ in $\mathbb{R}^n$ whose i^{th} component is $\langle u_i | b_j \rangle = \frac{\sqrt{p_i}}{\sqrt{\Pr(B_j)}}$ if $u_i \in B_j$, else 0, where $\Pr(B_j) = \sum_{u_i \in B_j} p_i$ is the probability of the block $B_j \in \pi$. The disjointness of the partition blocks translates into the orthogonality of these normalized vectors, i.e., $\langle b_j | b_{j'} \rangle = \delta_{jj'}$. Then the density matrix $\rho(\pi)$ is the probability sum of the outer-products or projectors $|b_j\rangle \langle b_j|$:

$$\rho(\pi) = \sum_{j=1}^{m} \Pr(B_j) |b_j\rangle \langle b_j|.$$

Then the entries are simply $\rho(\pi)_{ik} = \sqrt{p_i p_k}$ if there is a B_j with $u_i, u_k \in B_j$, else 0. Then the non-zero entries $\rho(\pi)_{ik} \neq 0$ correspond to the indistinctions $(u_i, u_k) \in \text{indit}(\pi)$ and the zero entries correspond to the distinctions $(u_i, u_k) \in \text{dit}(\pi)$.

Since the non-zero off-diagonal elements of $\rho(\pi)$ represent pairs of elements that cohere together in a block, they correspond to the "coherences" [18, p. 302] in the density matrices of QM.

> For this reason, the off-diagonal terms of a density matrix ... are often called "quantum coherences" because they are responsible for the interference effects typical of quantum mechanics that are absent in classical dynamics. [4, p. 177]

Consider an example on $U = \{a, b, c, d\}$ where $\pi = \{\{a, d\}, \{b, c\}\}$ where the point probabilities are $p_a = \frac{1}{3}$, $p_b = \frac{1}{4}$, $p_c = \frac{1}{6}$, and $p_d = \frac{1}{4}$. Then the block probabilities are $\Pr(\{a, d\}) = \frac{7}{12}$ and $\Pr(\{b, c\}) = \frac{5}{12}$. The density matrix of the partition is then:

$$\rho(\pi) = \begin{bmatrix} 1/3 & 0 & 0 & \sqrt{1/12} \\ 0 & 1/4 & \sqrt{1/24} & 0 \\ 0 & \sqrt{1/24} & 1/6 & 0 \\ \sqrt{1/12} & 0 & 0 & 1/4 \end{bmatrix}.$$

The non-zero off-diagonal elements correspond to the (non-trivial) indistinctions $\{(a, d), (d, a), (b, c), (c, b)\}$; the self-pairs of the diagonal are always indistinctions since no entity can be distinguished from itself, i.e., no element can appear in two different blocks of a partition. The off-diagonal zeros correspond to the distinctions of π, i.e., $\text{dit}(\pi) = \{(a,b),(a,c),(b,d),(c,d),...\}$ (where the ellipsis just stands for the reversed ordered pairs). The eigenvalues of $\rho(\pi)$ are always the block probabilities $\Pr(B_j)$ for $j = 1,...,m$ with the remaining $n-m$ eigenvalues being 0.

The density matrix representation of a set partition with point probabilities translates directly to the general version ρ of a density matrix in Hilbert space. A density matrix is not only Hermitian but positive so it has an orthonormal set of basis vectors $|u_i\rangle$ with non-negative real eigenvalues λ_i such that $\sum_i \lambda_i = 1$ which thus can be interpreted as the probabilities. In the set version, a "pure state" is the indiscrete partition $\mathbf{0}_U$ with single block probability of 1 and the quantum notion of a pure state density matrix, $\rho^2 = \rho$, has the only non-zero eigenvalue of 1. Hence we have a dictionary Table 5.1 giving the partition math and Hilbert space math versions of the notion of a quantum state.

Set level density matrix concept	Hilbert space density matrix concept
Density matrix $\rho(\pi)$	Quantum state density matrix ρ
Non-zero off-diagonal entries	Non-zero off-diagonal coherences of ρ
$\{\Pr(B_j)\}_{j=1}^m = $ non-0 eigenvalues	$\{\lambda_j\}_{j=1}^m$ eigenval., $\lambda_j \geq 0, \sum_{j=1}^m \lambda_j = 1$
$\rho(\pi) = \sum_{j=1}^m \Pr(B_j) |b_j\rangle \langle b_j|$	$\rho = \sum_{j=1}^m \lambda_j |u_j\rangle \langle u_j|$
Eigenvalue of 1 for $\rho(\mathbf{0}_U)$	Pure state $\rho = |u_j\rangle \langle u_j|$ has eigenvalue 1

Table 5.1: Set level and quantum state density matrices

6.6   Partitions and quantum observables

The observables of QM math do not carry probability information so the set level version of an observable will be a partition on a set without point probabilities but with values assigned to each block, i.e., the inverse image partitions $\{f^{-1}(r)\}_{r \in f(U)}$ of real-valued numerical attributes $f : U \to \mathbb{R}$. In the chapter on the quantum logic of vector space partitions (i.e., DSDs), the Yoga of Linearization was developed that produced the vector space concepts corresponding to set concepts (Table 4.1). In particular, every real-valued numerical attribute $f : U \to \mathbb{R}$ determines a quantum observable $F : V \to V$ defined on the basis set U by $Fu = f(u)u$. Here we develop a few more entries in the dictionary translating set concepts into Hilbert-space concepts in Table 5.2. Let the statement "$f \restriction S = rS$" mean that the numerical attribute $f : U \to \mathbb{R}$ is constant on the subset $S \subseteq U$ with value r. That is the set concept corresponding to the eigenvalue/eigenvector equation $Fu_i = f(u_i)u_i$ when U is interpreted as an ON basis for the Hilbert space. Thus a constant set S of f and value r correspond to an eigenvector and eigenvalue of $F : V \to V$. A characteristic (or indicator) function is a numerical attribute $\chi_S : U \to 2 = \{0, 1\}$ and it corresponds to a projection operator defined by $P_{[S]} u_i = \chi_S (u_i) u_i$ which only has eigenvalues of 0 and 1. The spectral decomposition of F in terms of the projectors $F = \sum_{r \in f(U)} r P_{V_r}$ then provides a 'spectral decomposition' $f = \sum_{r \in f(U)} r \chi_{f^{-1}(r)}$ of the numerical attribute f. Then the resolution of unity $\sum_{r \in f(U)} P_{V_r} = I : V \to V$ in the Hilbert-space case corresponds to the set version: $\cup_{r \in f(U)} \left(f^{-1}(r) \cap () \right) = I : \wp (U) \to \wp (U)$.

Set concept	Hilbert-space concept
Partition $\{f^{-1}(r)\}_{r \in f(U)}$	DSD $\{V_r\}_{r \in f(U)}$
$U = \uplus_{r \in f(U)} f^{-1}(r)$	$V = \oplus_{r \in f(U)} V_r$
Numerical attribute $f : U \to \mathbb{R}$	Observable $F u_i = f (u_i) u_i$
$f \upharpoonright S = rS$	$F u_i = r u_i$
Constant set S of f	Eigenvector u_i of F
Value r on constant set S	Eigenvalue r of eigenvector u_i
Characteristic fcn. $\chi_S : U \to \{0, 1\}$	Projection operator $P_{[S]} u_i = \chi_S (u_i) u_i$
Spectral Decomp. $f = \sum_{r \in f(U)} r \chi_{f^{-1}(r)}$	Spectral Decomp. $F = \sum_{r \in f(U)} r P_{V_r}$
$\cup_{r \in f(U)} \left(f^{-1}(r) \cap () \right) = I : \wp(U) \to \wp(U)$	$\sum_{r \in f(U)} P_{V_r} = I : V \to V$

Table 5.2: Skeletal set-level concepts and the corresponding vector (Hilbert) space concepts

6.7 Partitions and projective measurement

At the set level, a partition π with point probabilities defines a density matrix $\rho (\pi)$ and a numerical attribute $g : U \to \mathbb{R}$ defines an inverse-image partition $\sigma = \{g^{-1}(r)\}_{r \in g(U)}$. Their Hilbert space correlates are a quantum state ρ and a quantum observable $G : V \to V$. A projective measurement applies the observable to the density matrix. What is the set level version of applying the partition σ defined by the numerical attribute g to the partition $\rho (\pi)$? In Hilbert space, the effect of projective measurement is described by the Lüders mixture operation [4, p. 279] which yields the mixed state density matrix $\hat{\rho}$ according to the formula: $\hat{\rho} = \sum_{r \in g(U)} P_{V_r} \rho P_{V_r}$, where V_r is the eigenspace for the eigenvalue r of G, P_{V_r} is the projection operator to that subspace, and $g : U \to \mathbb{R}$ is the function assigning the eigenvalue to each u_i in the ON basis set of eigenvectors of G. The set version of the Lüders mixture operation is:

$$\hat{\rho} (\pi) = \sum_{r \in g(U)} P_{g^{-1}(r)} \rho (\pi) P_{g^{-1}(r)}$$

where the projection matrix $P_{g^{-1}(r)}$ is defined as the $n \times n$ diagonal matrix with the diagonal entries $\chi_{g^{-1}(r)} (u_i)$. Than it is easily shown that:

$$\hat{\rho}(\pi) = \rho(\pi \vee \sigma).$$

Thus the set level 'projective measurement' is just the partition join operation where $\text{dit}(\pi \vee \sigma) = \text{dit}(\pi) \cup \text{dit}(\sigma)$ and $\text{indit}(\pi \vee) = \text{indit}(\pi) \cap \text{indit}(\sigma)$.

For instance, if $\pi = \{\{a, b\}, \{c\}\}$ and $\sigma = \{\{a\}, \{b, c\}\}$, then $\pi \vee \sigma = \mathbf{1}_U$ as illustrated in Figure 6.6. The set level skeletonized version of a quantum jump is the jump from π to $\pi \vee \sigma = \mathbf{1}_U$ illustrated by the arrow in Figure 6.6.

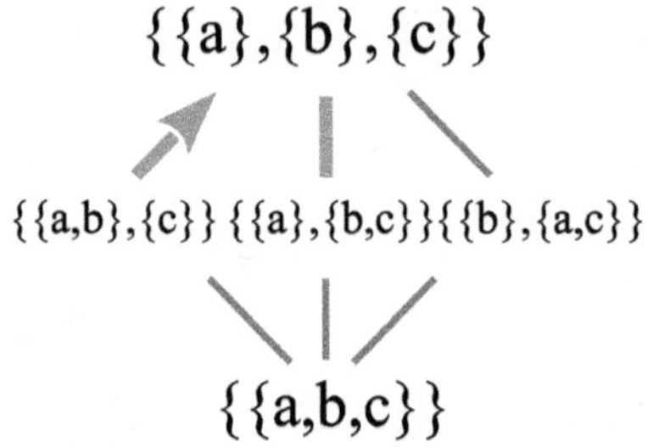

Figure 6.6: The set Lüders mixture operation yields the partition join

It may be useful to carry out the matrix operations. For $p_a = \frac{1}{2}$, $p_b = \frac{1}{6}$, and $p_c = \frac{1}{3}$, the density matrix $\rho(\pi)$ is:

$$\rho(\pi) = \begin{bmatrix} 1/2 & \sqrt{1/12} & 0 \\ \sqrt{1/12} & 1/6 & 0 \\ 0 & 0 & 1/3 \end{bmatrix}$$

and the Lüders mixture operation is:

$$\hat{\rho}(\pi) = \begin{bmatrix} 1 & 0 & 0 \\ 0 & 0 & 0 \\ 0 & 0 & 0 \end{bmatrix} \begin{bmatrix} \frac{1}{2} & \sqrt{\frac{1}{12}} & 0 \\ \sqrt{\frac{1}{12}} & \frac{1}{6} & 0 \\ 0 & 0 & \frac{1}{3} \end{bmatrix} \begin{bmatrix} 1 & 0 & 0 \\ 0 & 0 & 0 \\ 0 & 0 & 0 \end{bmatrix}$$

$$+ \begin{bmatrix} 0 & 0 & 0 \\ 0 & 1 & 0 \\ 0 & 0 & 1 \end{bmatrix} \begin{bmatrix} \frac{1}{2} & \sqrt{\frac{1}{12}} & 0 \\ \sqrt{\frac{1}{12}} & 1/6 & 0 \\ 0 & 0 & 1/3 \end{bmatrix} \begin{bmatrix} 0 & 0 & 0 \\ 0 & 1 & 0 \\ 0 & 0 & 1 \end{bmatrix}$$

$$= \begin{bmatrix} \frac{1}{2} & 0 & 0 \\ 0 & 0 & 0 \\ 0 & 0 & 0 \end{bmatrix} + \begin{bmatrix} 0 & 0 & 0 \\ 0 & \frac{1}{6} & 0 \\ 0 & 0 & \frac{1}{3} \end{bmatrix} = \begin{bmatrix} 1/2 & 0 & 0 \\ 0 & 1/6 & 0 \\ 0 & 0 & 1/3 \end{bmatrix} = \rho(\pi \vee \sigma) = \rho(\mathbf{1}_U).$$

This completes the dictionary entries for the set and Hilbert-space versions of quantum states, quantum observables, and projective measurement which are given in Table 5.3.

Partition math	Hilbert space math
State: $\rho(\pi) = \sum_{j=1}^{m} \Pr(B_j) \lvert b_j \rangle \langle b_j \rvert$	$\rho = \sum_{i=1}^{n} \lambda_i \lvert u_i \rangle \langle u_i \rvert$
Observable: $g = \sum_{r \in g(U)} r \chi_{g^{-1}(r)} : U \to \mathbb{R}$	$G = \sum_{r \in g(U)} r P_{V_r}$
Measurement: $\hat{\rho}(\pi) = \sum_{r \in g(U)} P_{g^{-1}(r)} \rho(\pi) P_{g^{-1}(r)}$	$\hat{\rho} = \sum_{r \in g(U)} P_{V_r} \rho P_{V_r}$

Table 5.3: Three basic notions: set version and corresponding Hilbert space version

6.8 Other aspects of QM mathematics

The partitional treatment of commuting, non-commuting, and conjugate observables was already treated in the chapter on the logic of DSDs. Since commuting observables correspond to numerical attributes defined on the same set, there is a set version of Dirac's Complete Set of Commuting Observables (CSCOs) to add to our dictionary.

- **Partition math**: A set of compatible partitions $\pi, \sigma, ..., \gamma$ defined by $f, g, ..., h : U \to \mathbb{R}$ is said to be *complete*, i.e., a Complete Set of Compatible Attributes or CSCA, if their join is the partition whose blocks are of cardinality one (i.e., $\mathbf{1}_U$). Then the elements $u \in U$ are uniquely characterized by the ordered set of values $(f(u), g(u), ..., h(u))$.

- **QM math**: A set of commuting observables $F, G, ..., H$ is said to be *complete*, i.e., a Complete Set of Commuting Observables or CSCO [21, 57], if the join of their eigenspace DSDs is the DSD whose subspaces are of dimension one. Then the simultaneous eigenvectors of the operators are unique characterized by the ordered set of their eigenvalues.

John von Neumann [108] divided quantum processes into two types: Type I (quantum measurements) and Type II (quantum evolution of an isolated system described by the Schrödinger equation). We have seen that the Type I measurement process is one of making distinctions, e.g., at the set level, joining the partition σ to the partition π to create $\pi \vee \sigma$ with the additional distinctions given by $\mathrm{dit}(\sigma) - \mathrm{dit}(\pi)$ since $\mathrm{dit}(\pi \vee \sigma) = \mathrm{dit}(\pi) \cup \mathrm{dit}(\sigma)$. Hence the natural characterization of a Type II process would be one that makes no distinctions between quantum states. Mathematically, the distinctness or rather indistinctness between two quantum states is given by their inner product (representing the overlap between the states). Thus zero indistinctness between two states (i.e., they are completely distinct with no overlap) is given by an inner product of zero, i.e., orthogonality. Hence a quantum process that makes no distinctions or indistinctions would be one that preserves inner products, i.e., does not change the overlap between quantum states. Such a transformation is a unitary transformation and it does characterize the Type II processes.[6] Hence the two types of quantum processes, those that create distinctions and those that don't can be illustrated at the skeletal set level as in Figure 6.7. At the skeletal level, a unitary process is just a non-singular transformation that transforms a basis set $U = \{a, b, c\}$ into another basis $U' = \{a', b', c'\}$.

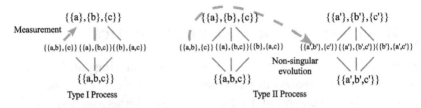

Figure 6.7: Skeletal version of Type I and Type II processes

It is particularly important to understand that there are different levels of indefiniteness, levels that are represented in the skeletal pictures of the partition lattices as being the quantum world 'beneath' the classical level represented by 1_U. Unitary evolution, which can exhibit the characteristic quantum interference effects, can take place at the non-classical level as illustrated in Figure 5.7. In the well-known double-slit experiment, the most perplexing case is when there is no detection at the slits (i.e., no distinguishing between the two slits) and a single particle will unitarily evolve showing the interference effects. Our biologically-evolved intuitions based on macroscopic reality assume a fully definite classical reality (represented by the disjoint partition 1_U in the partition lattice) so we intuitively assume that the particle in the superposition state |slit 1⟩ + |slit 2⟩ must first jump to the classical level before further evolving. But that is where our classical intuitions go wrong by insisting on an answer to the question: "Which slit did the particle go through?". As illustrated in Figure 5.7, a superposition can evolve at a non-classical level of indefiniteness. Hence there is no physical matter-of-fact of the particle going through one slit or the other, which is an answer resisted by our classical intuitions. This is an example of how the objective indefiniteness interpretation explains the experimental facts of the double-slit experiment without resorting to our classical

---

[6]The connection to solutions to the Schrödinger equation is given by Stone's Theorem [103].

intuitions (unlike the treatment in, say, Bohmian mechanics which appeals to ordinary intuitions of the particle going through one slit or the other [77, p. 144]).

A question that has been perplexing to quantum theorists and philosophers is the demarcation between Type I and Type II processes which we have seen is the making of distinctions or not. This answer is not new: it was in effect given by Richard Feynman as long ago as 1951 [38] and repeated many times in the Feynman Rules.

> If you could, *in principle*, distinguish the alternative *final* states (even though you do not bother to do so), the total, final probability is obtained by calculating the *probability* for each state (not the amplitude) and then adding them together. If you *cannot* distinguish the final states *even in principle*, then the probability amplitudes must be summed before taking the absolute square to find the actual probability.[40, 3-9]

Since at the logical level, information *is* distinctions [32], Anton Zeilinger makes the same point in terms of information.

> In other words, the superposition of amplitudes ... is only valid if there is no way to know, even in principle, which path the particle took. It is important to realize that this does not imply that an observer actually takes note of what happens. It is sufficient to destroy the interference pattern, if the path information is accessible in principle from the experiment or even if it is dispersed in the environment and beyond any technical possibility to be recovered, but in principle still "out there." The absence of any such information is the essential criterion for quantum interference to appear. [116, 484]

Figure 6.8 gives a simple graphic for the two Feynman cases to compute the probability of the particle to get from position A to position B. The particle can be thought of as piece of soft dough that is 'round' as a superposition of the four distinct polygonal shapes. The particle goes through an interaction with a grating. The measurement-or-no-measurement question is does the grating distinguish between the four shapes in the superposition or not.

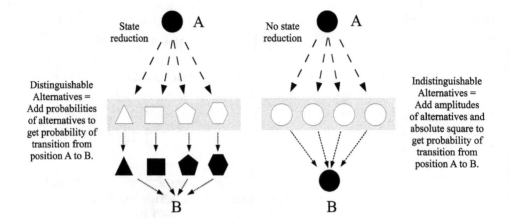

Figure 6.8: Two cases of a grating making distinctions or not

Feynman gives his own examples of an interaction where particle scatters off the atoms in a crystal. If there is no physical record of which atom the particle scattered off of, then one should add the amplitudes from scattering off the different atoms to get to B and than take the absolute square to get the probability of going from the original point A to B. But if all the atoms had, say, spin up, the particle had spin down, and the interaction reversed spin, then there is a physical record of which atom was scattered off of (its spin is reversed) to one should then add the probabilities of scattering off the different atoms to get from A to B.

It is also important to note that Feynman's treatment of "measurement" or (to use a less anthropomorphic term) state reduction takes place at the quantum level. For humans to observe the state reduction or not would require some means to magnify the result to the level of human observation, but those macroscopic components should play no role in the quantum theory. This treatment of state reduction at the quantum level thus bypasses the whole rather tortured literature of human level measurement as requiring an interaction with a macroscopic apparatus. This also bypasses the Zeh/Zurek notion of "decoherence" [117] involving the interaction with a macroscopic apparatus.

Chapter 7

Conclusion

From the mathematical viewpoint, subsets (or more generally, subobjects) are dual to partitions (or equivalence relations or quotient sets, or, more generally, quotient objects). Yet the historical development of "logic" has started with the logic of subsets (or the special case of propositional logic) and its extensions, variations, and sublogics. It has more or less completely ignored the logic of the dual concept of partitions. This book tries to repair that omission. In spite of the dual relationship, partitions are considerably more complex than subsets and this is reflected in the logic. Hence the results presented here and elsewhere ([24], [25]) are necessarily only a start.

In spite of the fragmentary development of partition logic, two major applications are already apparent. Gian-Carlo Rota set the stage for the first application with his comparison: $\frac{\text{Probability}}{\text{Subsets}} \approx \frac{\text{Information}}{\text{Partitions}}$. We saw that there is a thorough parallelism or duality between the notions of "elements of a subset" and "distinctions of a partition." Since Boole's logic of subsets naturally lead to the quantitative notion of the logical Boole-Laplace finite probability as the normalized number of elements in a subset (event) [13], it was natural to define the logical notion of information, logical entropy, as the normalized number of distinctions in a partition. The general formula for logical entropy, $h(\pi) = 1 - \sum_{j=1}^{m} \Pr(B_j)^2$ was not at all new but the partitional treatment, which defined information as distinctions, allowed a broader development providing a new foundation for information theory [32].

The concept of a partition or equivalence relation is the natural logical notion to define indefiniteness (pairs of elements in the same block or equivalence class) versus definiteness (pairs of elements in different blocks or equivalence classes) as well as the related concepts of indistinguishability versus distinguishability or identity versus difference. The notion of indefiniteness or indeterminacy has long been a standard part of quantum mechanics. For instance, a superposition of eigenstates in the observable basis is standardly considered not to have a definite value prior to some measurement-event to reduce the superposition to a definite eigenstate. The indefiniteness is highlighted by Heisenberg's Indeterminacy Principle for conjugate observables where the more definite is the value of one variable implies more indefiniteness in the value of the conjugate variable. Heisenberg and others such as Abner Shimony extrapolated those results in quantum mechanics to postulate a quantum world of real "potentialities" in addition to the classical world of actualities. The characteristic feature of the "potentialities" was their indeterminacy which only became determinate when "actualized." It is a vision of quantum reality as indefinite-world, not wave-world.

The second major application of the mathematics of partitions was to provide the proper logical framework to capture that vision of a quantum world of indeterminacy and indefiniteness. The mathematical framework of QM is distinctively different from that of classical physics. Hence our approach was to show that the distinctive mathematics (not the physics) of QM was the Hilbert space version of the mathematics of partitions. QM math differs from the math of classical physics in three basic notions, the quantum state, the quantum observable, and the projective measurement of a state by an observable. We showed that both the notions of state and observable were Hilbert space versions of different variations of partitions (i.e., a set partition with point probabilities and an inverse-image partition of a numerical attribute) and that a projective measurement as described by the Lüders mixture operation was prefigured at the partition level by the logical operation of the join of the two partitions. A number of other connections between partition math and QM math were made including the revealing overlap between this approach and the long-established Feynman rules based on distinguishability and indistinguishability.

The main point to emphasize is that it is early days for partition logic, both in the analysis and development of its complexities and in its applications–so much remains for future work.

Bibliography

[1] Ainsworth, Thomas. 2016. *Form vs. Matter*. In *The Stanford Encyclopedia of Philosophy (Spring 2016 Ed.)*, ed. Edward N. Zalta. https://plato.stanford.edu/archives/spr2016/entries/form-matter/.

[2] Ariew, Roger, ed. 2000. *G. W. Leibniz and Samuel Clarke: Correspondence*. Indianapolis: Hackett.

[3] Atkins, Peter, Julio de Paula, and James Keeler. 2018. *Atkins' Physical Chemistry 11th Ed.* Oxford UK: Oxford University Press.

[4] Auletta, Gennaro, Mauro Fortunato, and Giorgio Parisi. 2009. *Quantum Mechanics*. Cambridge UK: Cambridge University Press.

[5] Awodey, Steve. 2006. *Category Theory*. Oxford: Clarendon Press.

[6] Banaschewski, B. 1977. *On G. Spencer Brown's Laws of Form*. Notre Dame Journal of Formal Logic XVIII (3 July): 507–9.

[7] Baues, Hans-Joachim. 1995. *Homotopy Types*. In *Handbook of Algebraic Topology*, edited by I. M. James, 1–72. Amsterdam: Elsevier Science.

[8] Bennett, Charles H. 2003. Quantum Information: Qubits and Quantum Error Correction. *International Journal of Theoretical Physics* 42 (2 February): 153–76.

[9] Best, Michael J. 2017. *Quadratic Programming with Computer Programs*. Boca Raton FL: CRC Press.

[10] Birkhoff, Garrett 1948. *Lattice Theory*. New York: American Mathematical Society.

[11] Birkhoff, Garrett, and John Von Neumann. 1936. The Logic of Quantum Mechanics. *Annals of Mathematics* 37 (4): 823–43.

[12] Blyth, Thomas S. 2005. *Lattices and Ordered Algebraic Structures*. London: Springer-Verlag London.

[13] Boole, George 1854. *An Investigation of the Laws of Thought on which are founded the Mathematical Theories of Logic and Probabilities*. Cambridge: Macmillan and Co.

[14] Borceux, Francis. 1994. *Handbook of Categorical Algebra 3: Categories of Sheaves*. Cambridge UK: Cambridge University Press.

[15] Britz, Thomas, Matteo Mainetti, and Luigi Pezzoli 2001. Some operations on the family of equivalence relations. In *Algebraic Combinatorics and Computer Science: A Tribute to Gian-Carlo Rota*. H. Crapo and D. Senato eds., Milano: Springer: 445-59.

[16] Campbell, L. Lorne 1965. Entropy as a Measure. *IEEE Trans. on Information Theory*. IT-11 (January): 112-114.

[17] Church, Alonzo 1956. *Introduction to Mathematical Logic*. Princeton NJ: Princeton University Press.

[18] Cohen-Tannoudji, Claude, Bernard Diu, and Franck Laloë. 2005. *Quantum Mechanics: Volumes 1 and 2*. New York: John Wiley & Sons.

[19] Cover, Thomas and Joy Thomas 1991. *Elements of Information Theory*. New York: John Wiley.

[20] Csiszar, Imre, and Janos Körner. 1981. *Information Theory: Coding Theorems for Discrete Memoryless Systems*. New York: Academic Press.

[21] Dirac, P.A.M. 1958. *The Principles of Quantum Mechanics (4th ed.)*. Oxford: Clarendon Press.

[22] Dubreil, P. and M.-L. Dubreil-Jacotin 1939. Théorie algébrique des relations d'équivalence. *J. de Mathématique*. 18: 63-95.

[23] Eilenberg, Samuel, and Saunders Mac Lane. 1945. General Theory of Natural Equivalences. *Transactions of the American Mathematical Society* 58: 231–94.

[24] Ellerman, David 2010. The Logic of Partitions: Introduction to the Dual of the Logic of Subsets. *Review of Symbolic Logic*. 3 (2 June): 287-350.

[25] Ellerman, David. 2014. An Introduction of Partition Logic. *Logic Journal of the IGPL*. 22 (1): 94–125.

[26] Ellerman, David. 2017. Quantum mechanics over sets: a pedagogical model with non-commutative finite probability theory as its quantum probability calculus. *Synthese* 194: 4863–4896. https://doi.org/10.1007/s11229-016-1175-0.

[27] Ellerman, David. 2017. Logical Information Theory: New Foundations for Information Theory. *Logic Journal of the IGPL* 25 (5 Oct.): 806–35.

[28] Ellerman, David. 2018. The Quantum Logic of Direct-Sum Decompositions: The Dual to the Quantum Logic of Subspaces. *Logic Journal of the IGPL* 26 (1 January): 1–13. https://doi.org/10.1093/jigpal/jzx026.

[29] Ellerman, David. 2018. Logical Entropy: Introduction to Classical and Quantum Logical Information Theory. *Entropy* 20 (9): Article ID 679. https://doi.org/10.3390/e20090679.

[30] Ellerman, David. 2019. A Graph-Theoretic Method to Define Any Boolean Operation on Partitions. *The Art of Discrete and Applied Mathematics* 2 (2): 1–9. https://doi.org/10.26493/2590-9770.1259.9d5.

[31] Ellerman, David. 2021. The Logical Theory of Canonical Maps: The Elements & Distinctions Analysis of the Morphisms, Duality, Canonicity, and Universal Constructions in Set. *ArXiv.org*. https://arxiv.org/abs/2104.08583.

[32] Ellerman, David. 2021. *New Foundations for Information Theory: Logical Entropy and Shannon Entropy*. Cham, Switzerland: SpringerNature.

[33] Ellerman, David. 2021. *On Abstraction in Mathematics and Indefiniteness in Quantum Mechanics*. Journal of Philosophical Logic 50 (4): 813–35. https://doi.org/10.1007/s10992-020-09586-1.

[34] Ellerman, David. 2022. *Introduction to Logical Entropy and Its Relationship to Shannon Entropy*. 4Open Special Issue: Logical Entropy 5 (1): 1–33. https://doi.org/10.1051/fopen/2021004.

[35] Ellerman, David. 2022. *Follow the Math!: The Mathematics of Quantum Mechanics as the Mathematics of Set Partitions Linearized to (Hilbert) Vector Spaces*. Foundations of Physics 52 (5). https://doi.org/10.1007/s10701-022-00608-3.

[36] Ellerman, David. 2023. *Partitions, Objective Indefiniteness, and Quantum Reality: Towards the Objective Indefiniteness Interpretation of Quantum*

Mechanics. International Journal of Quantum Foundations 9 (2): 64–107. https://ijqf.org/archives/6789.

[37] Ellerman, David. (forthcoming). *Partitions, Indefiniteness, and Quantum Reality: The Objective Indefiniteness Interpretation of Quantum Mechanics.* Springer Briefs in Philosophy. Cham, Switzerland: Springer Nature.

[38] Feynman, Richard P. 1951. *The Concept of Probability in Quantum Mechanics.* In , 533–41. *Second Berkeley Symposium on Mathematical Statistics and Probability.* University of California Press.

[39] Feynman, Richard P., Robert B. Leighton, and Matthew Sands. 1963. *The Feynman Lectures on Physics: Mainly Mechanics, Radiation, and Heat Vol. I.* Reading MA: Addison-Wesley.

[40] Feynman, Richard P., Robert B. Leighton, and Matthew Sands. 1965. *The Feynman Lectures on Physics: Quantum Mechanics Vol. III.* Reading MA: Addison-Wesley.

[41] Finberg, David, Matteo Mainetti and Gian-Carlo Rota 1996. The Logic of Commuting Equivalence Relations. In *Logic and Algebra.* Aldo Ursini and Paolo Agliano ed., New York: Marcel Dekker: 69-96.

[42] Friedman, William F. 1922. *The Index of Coincidence and Its Applications in Cryptography.* Geneva IL: Riverbank Laboratories.

[43] Friedman, William F. 1986. *The Index of Coincidence and Its Applications in Cryptanalysis.* Walnut Creek CA: Aegean Park Press.

[44] Gini, Corrado 1912. *Variabilità e mutabilità.* Bologna: Tipografia di Paolo Cuppini.

[45] Gini, Corrado 1955. Variabilità e mutabilità. In *Memorie di metodologica statistica.* E. Pizetti and T. Salvemini eds., Rome: Libreria Eredi Virgilio Veschi.

[46] Gödel, Kurt. 1933. Zur intuitionistischen Arithmetik und Zahlentheorie. *Ergebnisse eines mathematischen Kolloquiums* 4: 34–38.

[47] Good, I. J. 1979. A.M. Turing's statistical work in World War II. *Biometrika.* 66 (2): 393-6.

[48] Good, I. J. 1982. Comment (on Patil and Taillie: Diversity as a Concept and its Measurement). *Journal of the American Statistical Association.* 77 (379): 561-3.

[49] Grätzer, George 2003. *General Lattice Theory* (2nd ed.). Boston: Birkhäuser Verlag.

[50] Halmos, Paul R. 1974. *Measure Theory*. New York: Springer-Verlag.

[51] Havrda, J. H. and F. Charvat 1967. Quantification Methods of Classification Processes: Concept of Structural α-Entropy. *Kybernetika (Prague)*. 3: 30-35.

[52] Heisenberg, Werner. 1962. *Physics & Philosophy: The Revolution in Modern Science*. New York: Harper Torchbooks.

[53] Hirschman, Albert O. 1945. *National power and the structure of foreign trade*. Berkeley: University of California Press.

[54] Hoffman, K., and R. Kunze. 1961. *Linear Algebra*. Englewood Cliffs NJ: Prentice-Hall.

[55] Hughes, R. I. G. 1989. *The Structure and Interpretation of Quantum Mechanics*. Cambridge: Harvard University Press.

[56] Jacobson, Nathan. 1953. *Lectures in Abstract Algebra Vol. II: Linear Algebra*. New York: Springer Science+Business Media.

[57] Jaynes, Edwin T. 1978. *Where Do We Stand on Maximum Entropy?* In *The Maximum Entropy Formalism*, edited by Raphael D. Levine and Myron Tribus, 15–118. Cambridge MA: MIT Press.

[58] Jaynes, Edwin T. 2003. *Probability Theory: The Logic of Science*. Edited by G. Larry Bretthorst. Cambridge UK: Cambridge University Press.

[59] Kahn, David 1967. *The Codebreakers: The Story of Secret Writing*. New York: Macmillan.

[60] Kant, Immanuel. 1998. *Critique of Pure Reason*. Translated by Paul Guyer and Allen W. Wood. The Cambridge Edition of the Works of Immanuel Kant. Cambridge UK: Cambridge University Press.

[61] Kaplan, Wilfred. 1999. *Maxima and Minima with Applications: Practical Optimization and Duality*. New York: John Wiley & Sons.

[62] Kastner, Ruth E. 2013. *The Transactional Interpretation of Quantum Mechanics: The Reality of Possibility*. New York: Cambridge University Press.

[63] Kendall, M. G. 1945. *Advanced Theory of Statistics Vol. I.* London: Charles Griffin & Co.

[64] Kolmogorov, Andrei N. 1983. Combinatorial Foundations of Information Theory and the Calculus of Probabilities. *Russian Math. Surveys* 38 (4): 29–40.

[65] Kullback, Solomon 1976. *Statistical Methods in Cryptanalysis.* Walnut Creek CA: Aegean Park Press.

[66] Kung, Joseph P. S., Gian-Carlo Rota and Catherine H. Yan 2009. *Combinatorics: The Rota Way.* New York: Cambridge University Press.

[67] Lawvere, F. William. 1986. Introduction. In *Categories in Continuum Physics (Buffalo 1982) LNM 1174*, ed. F. William Lawvere and Stephen Schanuel, 1–16. Springer-Verlag.

[68] Lawvere, F. William. 1991. Intrinsic co-Heyting boundaries and the Leibniz rule in certain toposes. In *Category Theory (Como 1990) LNM 1488*, ed. A. Carboni, M. C. Pedicchio, and G. Rosolini, 279–297. Springer-Verlag.

[69] Lawvere, F. William, and Robert Rosebrugh. 2003. *Sets for Mathematics.* Cambridge MA: Cambridge University Press.

[70] Leibniz, G. W. 1996. *New Essays on Human Understanding.* Translated by Peter Remnant and Jonathan Bennett. Cambridge UK: Cambridge University Press.

[71] Lewis, Peter J. 2016. *Quantum Ontology: A Guide to the Metaphysics of Quantum Mechanics.* New York: Oxford University Press.

[72] Mac Lane, Saunders. 1948. Groups, categories, and duality. *Proc. Nat. Acad. Sci. U.S.A.* 34: 263–67.

[73] Mac Lane, Saunders. 1998. *Categories for the Working Mathematician. 2nd Ed.* New York: Springer Science+Business Media.

[74] Mac Lane, Saunders, and Ieke Moerdijk. 1992. *Sheaves in Geometry and Logic: A First Introduction to Topos Theory.* New York: Springer.

[75] MacKay, D. J. C. 2003. *Information Theory, Inference, and Learning Algorithms.* Cambridge UK: Cambridge University Press.

[76] Margenau, Henry. 1954. *Advantages and Disadvantages of Various Inter-pretations of the Quantum Theory. Physics Today* 7 (10): 6–13.

[77] Maudlin, Tim. 2019. *Philosophy of Physics: Quantum Theory.* Princeton NJ: Princeton University Press.

[78] Mermin, David. 2012. *Commentary: Quantum Mechanics: Fixing the Shifty Split. Physics Today* 65 (7): 8–10. https://doi.org/10.1063/pt.3.1618.

[79] Ore, Oystein 1942. Theory of equivalence relations. *Duke Mathematical Journal.* 9: 573-627.

[80] Pagels, Heinz. 1985. *Perfect Symmetry: The Search for the Beginning of Time.* New York: Simon and Schuster.

[81] Patil, G. P. and C. Taillie 1982. Diversity as a Concept and its Measurement. *Journal of the American Statistical Association.* 77 (379): 548-61.

[82] Peirce, Charles S. 1880. On the Algebra of Logic. *American Journal of Mathematics* 13 (1): 15–57.

[83] Rao, C. Radhakrishna 1982. Diversity and Dissimilarity Coefficients: A Unified Approach. *Theoretical Population Biology.* 21: 24-43.

[84] Rao, C. R. 1982. *Gini-Simpson Index of Diversity: A Characterization, Generalization and Applications. Utilitas Mathematica B* 21: 273–82.

[85] Rao, C. R. 2010. *Quadratic Entropy and Analysis of Diversity. Sankhyā: The Indian Journal of Statistics* 72-A (1): 70–80.

[86] Rasiowa, Helena. 1974. *An Algebraic Approach to Non-Classical Logics.* Amsterdam: North-Holland.

[87] Rauszer, Cecylia. 1974. A Formalization of the Propositional Calculus of H-B Logic. *Studia Logica.* 33: 23–34.

[88] Rejewski, M. 1981. How Polish Mathematicians Deciphered the Enigma. *Annals of the History of Computing.* 3: 213-34.

[89] Ricotta, Carlo and Laszlo Szeidl 2006. Towards a unifying approach to diversity measures: Bridging the gap between the Shannon entropy and Rao's quadratic index. *Theoretical Population Biology.* 70: 237-43.

[90] Rota, Gian-Carlo. 1964. The number of partitions of a set. *American Mathematical Monthly* 71: 498–504. https://doi.org/10.2307/2312585.

[91] Rota, Gian-Carlo. 1997. *Indiscrete Thoughts*. Boston: Birkhäuser.

[92] Rota, Gian-Carlo. 2001. Twelve Problems in Probability No One Likes to Bring up. In *Algebraic Combinatorics and Computer Science: A Tribute to Gian-Carlo Rota*, edited by Henry Crapo and Domenico Senato, 57–93. Milano: Springer.

[93] Ryser, Herbert John. 1963. *Combinatorial Mathematics*. Washington DC: Mathematical Association of America.

[94] Shannon, Claude E. 1948. A Mathematical Theory of Communication. *Bell System Technical Journal*. 27: 379-423; 623-56.

[95] Shannon, Claude E. and Warren Weaver 1964. *The Mathematical Theory of Communication*. Urbana: University of Illinois Press.

[96] Shannon, Claude E. 1993. *The Bandwagon*. In *Claude E. Shannon: Collected Papers*, edited by N. J. A. Sloane and Aaron D. Wyner, 462. Piscataway NJ: IEEE Press.

[97] Shannon, Claude E. 1993. *Some Topics in Information Theory*. In *Claude E. Shannon: Collected Papers*, edited by N. J. A. Sloane and Aaron D. Wyner, 458–59. Piscataway NJ: IEEE Press.

[98] Shimony, Abner. 1988. *The Reality of the Quantum World*. Scientific American 258 (1): 46–53.

[99] Shimony, Abner. 1999. *Philosophical and Experimental Perspectives on Quantum Physics*. In *Philosophical and Experimental Perspectives on Quantum Physics: Vienna Circle Institute Yearbook* 7, 1–18. Dordrecht: Springer Science+Business Media.

[100] Simpson, Edward Hugh 1949. Measurement of Diversity. *Nature*. 163: 688.

[101] Sinkov, Abraham 1968. *Elementary Cryptanalysis: A Mathematical Approach*. New York: Random House.

[102] Spenser Brown, G. 1972. *Laws of Form*. New York: The Julian Press.

[103] Stone, M. H. 1932. *On One-Parameter Unitary Groups in Hilbert Space*. *Annals of Mathematics* 33 (3): 643–48.

[104] Tamir, Boaz, Ismael Lucas Piava, Zohar Schwartzman-Nowik, and Eliahu Cohen. 2022. *Quantum Logical Entropy: Fundamentals and General Properties. 4Open Special Issue: Logical Entropy* 5 (2): 1–14. https://doi.org/10.1051/fopen/2021005.

[105] Tsallis, C. 1988. Possible Generalization for Boltzmann-Gibbs statistics. *J. Stat. Physics.* 52: 479-87.

[106] Univalent Foundations Program. 2013. *Homotopy Type Theory: Univalent Foundations of Mathematics.* Institute for Advanced Studies, Princeton. http://homotopytypetheory.org/book.

[107] Vajda, I. 1969. A Contribution to Informational Analysis of Patterns. In *Methodologies of Pattern Recognition.* Satosi Watanabe ed., New York: Academic Press: 509-519.

[108] Von Neumann, John. 1955. *Mathematical Foundations of Quantum Mechanics.* Translated by Robert Beyer. Princeton: Princeton University Press.

[109] Weinberg, Steven. 2014. *Quantum Mechanics without State Vectors. Physical Review A* 90: 042102. https://doi.org/10.1103/PhysRevA.90.042102.

[110] Weinberg, Steven. 2015. *Lectures on Quantum Mechanics 2nd Ed.* Cambridge UK: Cambridge University Press.

[111] Weyl, Hermann. 1949. *Philosophy of Mathematics and Natural Science.* Princeton: Princeton University Press.

[112] Whitman, P.M. 1946. Lattices, equivalence relations, and subgroups. *Bull. Am. Math. Soc.* 52: 507-522.

[113] Wilson, Robin J. 1972. *Introduction to Graph Theory.* London: Longman.

[114] Wolter, Frank. 1998. On logics with coimplication. *Journal of Philosophical Logic.* 27: 353–387.

[115] Yeung, Raymond W. 2002. *A First Course in Information Theory.* New York: Springer Science+Business Media.

[116] Zeilinger, Anton. 1999. *Experiment and the Foundations of Quantum Physics.* In *More Things in Heaven and Earth,* edited by B. Bederson, 482–98. Berlin: Springer-Verlag.

[117] Zurek, Wojciech Hubert. 2003. *Decoherence, Einselection, and the Quantum Origins of the Classical. Review of Modern Physics* 75 (3 July): 715–75.

Index

www.ingramcontent.com/pod-product-compliance
Lightning Source LLC
Chambersburg PA
CBHW071120050326
40690CB00008B/1280